T0230470

Lecture Notes in Physics

Founding Editors

Wolf Beiglböck

Jürgen Ehlers

Klaus Hepp

Hans-Arwed Weidenmüller

Volume 1019

Series Editors

Roberta Citro, Salerno, Italy

Peter Hänggi, Augsburg, Germany

Morten Hjorth-Jensen, Oslo, Norway

Maciej Lewenstein, Barcelona, Spain

Luciano Rezzolla, Frankfurt am Main, Germany

Angel Rubio, Hamburg, Germany

Wolfgang Schleich, Ulm, Germany

Stefan Theisen, Potsdam, Germany

James D. Wells, Ann Arbor, MI, USA

Gary P. Zank, Huntsville, AL, USA

The series Lecture Notes in Physics (LNP), founded in 1969, reports new developments in physics research and teaching - quickly and informally, but with a high quality and the explicit aim to summarize and communicate current knowledge in an accessible way. Books published in this series are conceived as bridging material between advanced graduate textbooks and the forefront of research and to serve three purposes:

- to be a compact and modern up-to-date source of reference on a well-defined topic;
- to serve as an accessible introduction to the field to postgraduate students and non-specialist researchers from related areas;
- to be a source of advanced teaching material for specialized seminars, courses and schools.

Both monographs and multi-author volumes will be considered for publication. Edited volumes should however consist of a very limited number of contributions only. Proceedings will not be considered for LNP.

Volumes published in LNP are disseminated both in print and in electronic formats, the electronic archive being available at springerlink.com. The series content is indexed, abstracted and referenced by many abstracting and information services, bibliographic networks, subscription agencies, library networks, and consortia.

Proposals should be sent to a member of the Editorial Board, or directly to the responsible editor at Springer:

Dr Lisa Scalone
lisa.scalone@springernature.com

Omar Benhar

Structure and Dynamics of Compact Stars

 Springer

Omar Benhar
Physics
INFN and Sapienza University of Rome
Rome, Italy

ISSN 0075-8450   ISSN 1616-6361   (electronic)
Lecture Notes in Physics
ISBN 978-3-031-35627-8   ISBN 978-3-031-35628-5   (eBook)
https://doi.org/10.1007/978-3-031-35628-5

© The Editor(s) (if applicable) and The Author(s), under exclusive license to Springer Nature Switzerland
AG 2023
This work is subject to copyright. All rights are solely and exclusively licensed by the Publisher, whether
the whole or part of the material is concerned, specifically the rights of translation, reprinting, reuse
of illustrations, recitation, broadcasting, reproduction on microfilms or in any other physical way, and
transmission or information storage and retrieval, electronic adaptation, computer software, or by similar
or dissimilar methodology now known or hereafter developed.
The use of general descriptive names, registered names, trademarks, service marks, etc. in this publication
does not imply, even in the absence of a specific statement, that such names are exempt from the relevant
protective laws and regulations and therefore free for general use.
The publisher, the authors, and the editors are safe to assume that the advice and information in this book
are believed to be true and accurate at the date of publication. Neither the publisher nor the authors or
the editors give a warranty, expressed or implied, with respect to the material contained herein or for any
errors or omissions that may have been made. The publisher remains neutral with regard to jurisdictional
claims in published maps and institutional affiliations.

This Springer imprint is published by the registered company Springer Nature Switzerland AG
The registered company address is: Gewerbestrasse 11, 6330 Cham, Switzerland

Paper in this product is recyclable.

Preface

White dwarfs and neutron stars, collectively referred to as compact stars, are believed to be the final stages of evolution of massive stars. They are very different from ordinary stars, whose stability against gravitational collapse is due to thermal pressure. In compact stars, the nuclear fuel necessary to ignite the fusion reactions leading to heat production is no longer available, and the pressure required for hydrostatic equilibrium originates from different mechanisms, driven by quantum-mechanical effects and interactions between the constituents of matter in the star interior.

Compact stars provide a unique laboratory to study the structure of matter under extreme conditions not attainable on Earth, in which all known forces—gravity, electromagnetic interactions, as well as weak and strong nuclear interactions—play a fundamental role. As a consequence, studying the properties of these systems is relevant to a variety of research fields, ranging from astrophysics and gravitational physics to nuclear and particle physics.

Early efforts aimed at shedding light on the properties of matter at densities exceeding by many orders of magnitude the values typical of terrestrial macroscopic objects, $\varrho \lesssim 20\,\mathrm{g/cm^3}$, date back to the 1920s. These investigations were triggered by the discovery of the star named Sirius B, whose internal density was found to be as high as $3 \times 10^6\,\mathrm{g/cm^3}$. Sirius B is now known to belong to the class of stars called white dwarfs.

The momentous discovery of an upper bound to the mass of white dwarfs, made by Chandrasekhar in 1931, started a wave of speculations on the fate of stars of mass exceeding the limiting value. Depending on their mass, these stars can either evolve into neutron stars or give rise to the appearance of a black hole, that is, a region of space where the gravitational field is so intense that neither matter nor radiation can escape.

Neutron stars, whose properties are said to have been first discussed by Bohr, Landau, and Rosenfeld in the days immediately following the discovery of the neutron, in 1932, were predicted to appear in the aftermath of a supernova explosion by Baade and Zwicky. Owing to the large densities involved, up to and above $10^{15}\,\mathrm{g/cm^3}$, the description of neutron star matter involves a number of challenging issues, reflecting the complexity of microscopic dynamics at both nuclear and subnuclear level. If the density becomes much larger than the central density of atomic nuclei, $\varrho_0 \approx 2.6 \times 10^{14}\,\mathrm{g/cm^3}$, new forms of matter predicted by the fundamental theories of weak and strong interactions, such as strange baryonic matter and deconfined quark matter, are also expected to appear. In addition, the effects of the spacetime curvature described by Einstein's theory of general relativity—which turn out to be negligible in white dwarfs—play a primary role in the determination of neutron star properties.

This volume is based on the courses of *Physics of Dense Matter* and *Structure of Compact Stars* that I have been teaching to master and doctoral physics students at Sapienza University of Rome since 2005. The primary purpose of these classes was providing a concise and self-contained introduction to the structure of matter at densities $10^4 \lesssim \varrho \lesssim 10^{15}$ g/cm^3, typical of white dwarfs and neutron stars, to students with a broad range of interests. The discussion is mainly focused on the region of $\varrho > \varrho_0$—the nuclear density ϱ_0 being the largest density observed on Earth under ordinary conditions—whose description necessarily involves a significant amount of extrapolation of the available empirical knowledge of nuclear properties.

A special emphasis is given to the recent applications of nuclear matter theory to the analysis of astrophysical data, providing information on neutron star properties useful to constrain theoretical models of dense matter. A prominent role, in this context, is played by the observations of gravitational-wave emission from neutron stars, which arguably opened a whole new era for both astrophysics and nuclear physics research.

Over the years, I benefited from countless stimulating discussions with my fellows of INFN and Sapienza University, my collaborators, and my students. Their insightful comments and advice are gratefully acknowledged. A special mention is owed to my longtime friends and colleagues Stefano Fantoni and Valeria Ferrari, whose influence largely contributed to shape this book. Finally, I would like to emphasize how much I am indebted to my late friends and collaborators Adelchi Fabrocini, Vijay R. Pandharipande, and Artur Polls, who gave fundamental and lasting contributions to the understanding of the physics of dense matter and neutron stars.

Unless otherwise stated, throughout the book I use the system of natural units, in which $c = \hbar = k_B = 1$.

Roma, Italy   Omar Benhar
April 2023

Contents

Acronyms

AFDMC	Auxiliary-Field Diffusion Monte Carlo
APR	Akmal Pandharipande Ravenhall
BCC	Body-Centered Cubic
BHF	Brueckner Hartree Fock
CBF	Correlated Basis Functions
EOS	Equation of State
GBM	Gamma-Ray Burst Monitor
GRB	Gamma-Ray Burst
GFMC	Green Function Monte Carlo
GW	Gravitational Waves
INTEGRAL	International Gamma-Ray Astrophysics Laboratory
LIGO	Laser Interferometer Gravitational-Wave Observatory
NICER	Neutron Star Interior Composition Explorer
NJL	Nambu Jona–Lasinio
NN	Nucleon-Nucleon
NNN	Three-Nucleon
OPE	One Pion Exchange
QCD	Quantum Chromo Dynamics
QNM	Quasi-normal Modes
RMF	Relativistic Mean Field
SNM	Symmetric Nuclear Matter
SQM	Strange Quark Matter
PNM	Pure Neutron Matter
VMC	Variational Monte Carlo
XMM	X-ray Multi-mirror Mission
XTI	X-ray Timing Instrument

Physical and Astronomical Constants

Axial-vector coupling constant; $g_A = 1.26$
Boltzmann constant; $k_B = 8.617 \times 10^{-11}\,\mathrm{MeVK}^{-1}$
Cabibbo angle; $\theta_c = 13.02\ \mathrm{deg}$
Charged pion mass; $m_{\pi^\pm} = 139.6\ \mathrm{MeV}$
Electron mass; $m_e = 0.511\ \mathrm{MeV} = 9.11 \times 10^{-28}\ \mathrm{g}$
Nuclear matter equilibrium density; $\varrho_0 = 2.6 \times 10^{14}\ \mathrm{g\,cm}^{-3}$
Fermi constant; $G_F = 1.166 \times 10^{-5}\ \mathrm{GeV}^{-2}$
Gravitational constant; $G = 6.674 \times 10^{-8}\ \mathrm{cm}^3\,\mathrm{g}^{-1}\,\mathrm{s}^{-2} = 6.709 \times 10^{-39}\ \mathrm{GeV}^{-2}$
Muon mass; $m_\mu = 105.7\ \mathrm{MeV}$
Neutral pion mass; $m_{\pi^0} = 135\ \mathrm{MeV}$
Neutron mass; $m_n = 939.6\ \mathrm{MeV} = 1.67 \times 10^{-24}\ \mathrm{g}$
Nucleon mass (isospin average); $m = 939\ \mathrm{MeV}$
Omega-meson mass; $m_\omega = 783\ \mathrm{MeV}$
Pion mass (isospin average); $m_\pi = 138.1\ \mathrm{MeV}$
Pion-nucleon coupling constant; $g^2/4\pi = 13.5$
Proton mass; $m_p = 938.3\ \mathrm{MeV} = 1.67 \times 10^{-24}\ \mathrm{g}$
Rho-meson mass; $m_\rho = 769\ \mathrm{MeV}$
Solar Luminosity; $L_\odot = 3.89 \times 10^{33}\ \mathrm{erg\,s}^{-1}$
Solar mass; $M_\odot = 1.988 \times 10^{33}\ \mathrm{g}$
Solar radius; $R_\odot = 6.957 \times 10^5\ \mathrm{km}$
Stephan-Boltzmann constant; $\sigma = 5.67 \times 10^{-5}\ \mathrm{erg\,cm}^{-2}\,\mathrm{s}^{-1}\,\mathrm{K}^{-4}$

Part I

White Dwarfs

The Prototype Compact Star

1

Abstract

The discovery of the first white dwarf, a star with mass comparable to the solar mass and the size of a planet, revealed the existence of a new form of matter, of density exceeding that of terrestrial macroscopic objects by many orders of magnitude. This chapter provides an overview of white dwarf properties, highlighting the role played by the equation of state of matter in the star interior. The quantum origin of the pressure balancing the gravitational pull and the emergence of a limiting mass of stable white dwarfs, famously predicted by Chandrasekhar in the 1970s, are also discussed.

1.1 Discovery of White Dwarfs

In 1844, the German astronomer Friedrich Bessel deduced that the star called Sirius had a yet unseen companion, that was then observed two decades later by Alvan Graham Clark and named Sirius B. The mass of Sirius B was determined by applying Kepler's third law to the orbit of the binary system, while its radius was obtained in the 1920s from the equation describing blackbody emission, using the measured spectrum and luminosity. The resulting values, $M \sim 0.75$ to $0.95\ M_\odot$—with $M_\odot = 1.988 \times 10^{33}$ g being the mass of the Sun—and $R \sim 18,800$ km, comparable to the typical radius of a planet,[1] revealed that Sirius B was am extraordinarily compact object, with density reaching millions of g/cm^3. In his book *The Internal Constitution of the Stars*, published in 1926, Sir Arthur Eddington famously wrote: "we have a star of mass about equal to the Sun and radius much

---

[1] Note that this figure is about four times bigger than the value resulting from more recent measurements; see Table 1.2.

© The Author(s), under exclusive license to Springer Nature Switzerland AG 2023
O. Benhar, *Structure and Dynamics of Compact Stars*, Lecture Notes
in Physics 1019, https://doi.org/10.1007/978-3-031-35628-5_1

less than Uranus" [1]. Sirius B is now known to belong to the class of stellar objects
called white dwarfs.

1.2   Formation of White Dwarfs

The formation of a star is triggered by the contraction of a self-gravitating hydrogen
cloud. As the density increases, the cloud becomes more and more opaque, and
the energy released cannot be efficiently radiated away. As a consequence, the
temperature also increases, and eventually reaches the value, $T \sim 6 \times 10^7$ K, needed
to ignite the nuclear reactions turning hydrogen into helium

$$p + p \rightarrow {}^2\text{H} + e^+ + \nu + 0.4\,\text{MeV} ,$$

$$e^+ + e^- \rightarrow \gamma + 1.0\,\text{MeV} ,$$

$${}^2\text{H} + p \rightarrow {}^3\text{He} + \gamma + 5.5\,\text{MeV} ,$$

$${}^3\text{He} + {}^3\text{He} \rightarrow {}^4\text{He} + 2p + 26.7\,\text{MeV} .$$

Note that the above reactions are all *exothermic*, with the energy being released in
form of kinetic energy of the produced particles.[2] Equilibrium is reached as soon as
gravitational attraction is balanced by matter pressure.

When the nuclear fuel is exhausted, the core stops producing heat, the internal pressure cannot be sustained, and the contraction generated by gravitational
attraction resumes. If the mass of the helium core is large enough, this contraction,
associated with a further increase of the temperature, leads to the ignition of new
fusion reactions, resulting in the appearance of heavier nuclei, such as carbon and
oxygen. Depending on the mass of the progenitor star, this process can take place
several times, the final result being the formation of a core made of the most stable
nuclear species, nickel and iron, at density comparable to highest value observed
on Earth, $\varrho_0 \approx 2.67 \times 10^{14}\,\text{g/cm}^3$, typical of the central region of heavy atomic
nuclei. Even larger densities are believed to occur in the interior of neutron stars,
astrophysical objects resulting from the contraction of the iron core in very massive
stars, having $M > 4\,M_\odot$. Note that during the contraction the progenitor star loses
a large amount of matter from its outer layers, and the mass of the newly formed
compact star—be it a white dwarf or a neutron star—is determined by the mass
of the leftover. The evolutionary stages of a progenitor star of mass $\sim 25\,M_\odot$ are
summarised in Table 1.1 .

If the star is sufficiently small, so that the gravitational contraction of the core
does not produce a temperature high enough to ignite the burning of heavy nuclei, it
will eventually turn into a white dwarf, consisting primarily of helium, carbon and
oxygen.

---

[2] Recall: 1 MeV = 1.6021917×10^{-6} erg.

Table 1.1 Stages of nucleosynthesis for a star of mass $\sim 25\ M_\odot$. Adapted from Ref. [2] with permissions, © APS 2002. All rights reserved

Nuclear fuel	Main products	Temperature (K)	Density $(g\ cm^{-3})$	Duration (yrs)
H	He	3.81×10^7	3.81	6.70×10^6
He	C, O	1.96×10^8	762	8.39×10^5
C	O, Ne, Mg	8.41×10^8	1.29×10^5	522
Ne	O, Mg, Si	1.57×10^9	3.95×10^6	0.891
O	Si, S	2.09×10^9	3.60×10^6	0.402
Si	Fe	3.65×10^9	3.01×10^7	0.002

Table 1.2 Measured values of mass and radius of three white dwarfs, in units of the solar mass and radius $M_\odot$ and $R_\odot$

	Mass $(M_\odot)$	Radius $(R_\odot)$
Sirius B [3,4]	1.018 ± 0.011	0.0084 ± 0.0002
Procyon B [5]	0.602 ± 0.015	0.01234 ± 0.00032
40 Eri B [6,7]	0.573 ± 0.018	0.0136 ± 0.0002

The over 2000 observed white dwarfs have luminosity $L \sim 10^{-2}\ L_\odot$—with $L_\odot$ being the luminosity of the Sun—and surface temperature $T_s \sim 10^4$ K. The radius of many white dwarfs has been determined from their measured flux, defined as

$$F(D) = \frac{L}{4\pi D^2} \ , \tag{1.1}$$

where D is the distance from the Earth, obtained applying the parallax method. Combining the above equation to the equation describing black body emission

$$L = 4\pi R^2 \sigma T_s^4 \ , \tag{1.2}$$

where σ is the constant of Stephan-Boltzmann, one obtains

$$R = \sqrt{\frac{F D^2}{\sigma T_s^4}} \ . \tag{1.3}$$

As an example, the measured values of mass and radius of three white dwarfs are given in Table 1.2. The corresponding average densities are in the range 10^6–10^7 g/cm^3, to be compared to the typical density of terrestrial macroscopic objects, not exceeding 20 g/cm^3.

In his fundamental paper of 1931 [8], Chandrasekhar demonstrated that the pressure required to ensure the stability of white dwarfs against gravitational collapse is provided by a gas of noninteracting electrons at very low temperature. The main properties of this systems are reviewed in the following section.

1.3   Properties of the Degenerate Fermi Gas

Let us consider a system of noninteracting electrons uniformly distributed in a cubic box of volume $V = L^3$. If the temperature is low enough that thermal energies can be neglected, the lowest quantum levels are occupied by two electrons, one for each spin state. This configuration corresponds to the ground state of the system. A gas of noninteracting electrons in its ground state is said to be *fully degenerate*. At higher temperature, the thermal energy can excite electrons to higher energy states, leaving some of the lower lying levels empty. In this case, the system is no longer fully degenerate.

As the electrons are uniformly distributed, their wave functions exhibit translational invariance. They can be written in the form

$$\psi_{\mathbf{p}s}(\mathbf{r}) = \phi_{\mathbf{p}}(\mathbf{r})\chi_s \,, \tag{1.4}$$

where χ_s is a Pauli spinor specifying the spin projection and

$$\phi_{\mathbf{p}}(\mathbf{r}) = \sqrt{\frac{1}{V}} \, e^{i\mathbf{p}\cdot\mathbf{r}} \,, \tag{1.5}$$

is an eigenfunctions of the momentum operator—the generator of translations in space—satisfying the periodic boundary conditions

$$\phi_{\mathbf{p}}(x, y, z) = \phi_{\mathbf{p}}(x + n_x L, y + n_y L, z + n_z L) \,. \tag{1.6}$$

Here, x, y and z denote the components of the vector $\mathbf{r}$, specifying the electron position, and $n_x, n_y, n_z = 0, \pm 1, \pm 2, \ldots$. The above equation obviously implies the relations

$$p_x = \frac{2\pi n_x}{L}, \quad p_y = \frac{2\pi n_y}{L}, \quad p_z = \frac{2\pi n_z}{L} \,, \tag{1.7}$$

with $\mathbf{p} \equiv (p_x, p_y, p_z)$, which in turn determine the momentum eigenvalues.

Each quantum state is associated with an eigenvalue of the momentum $\mathbf{p}$, that is, with a specific triplet of integers (n_x, n_y, n_z). The corresponding energy eigenvalue turns out to be

$$\epsilon_p = \frac{p^2}{2m_e} = \left(\frac{2\pi}{L}\right)^2 \frac{1}{2m_e} \, (n_x^2 + n_y^2 + n_z^2) \,, \tag{1.8}$$

where $p^2 = |\mathbf{p}|^2 = p_x^2 + p_y^2 + p_z^2$, and m_e denotes the electron mass. The maximum electron energy in a fully degenerate gas is called Fermi Energy, and denoted by ϵ_F. The corresponding momentum, called Fermi momentum, is $p_F = \sqrt{2m_e\epsilon_F}$.[3]

---

[3] Here we consider a non relativistic system. The generalisation needed to describe relativistic electrons will be discussed at a later stage.

1.3.1   Energy Density

The number of quantum states with energy less or equal to ϵ_F can be easily calculated. Since each triplet (n_x, n_y, n_z) corresponds to a point in a cubic lattice with unit lattice spacing, the number of momentum eigenstates is equal to the number of lattice points within a sphere of radius $R = p_F L/(2\pi)$. The number of electrons in the system can then be obtained from

$$N = 2 \frac{4\pi}{3} R^3 = V \frac{p_F^3}{3\pi^2} , \qquad (1.9)$$

where the factor 2 takes into account spin degeneracy, implying that there are two electrons with opposite spin projections in each momentum eigenstate. It follows that the electron number density, that is, the number of electrons per unit volume, is given by

$$n_e = \frac{N}{V} = \frac{p_F^3}{3\pi^2} . \qquad (1.10)$$

The total ground state energy can now be easily evaluated from

$$E = 2 \sum_{p \leq p_F} \frac{p^2}{2m_e} , \qquad (1.11)$$

by using Eq. (1.7) and taking the limit $L \to \infty$, which corresponds to vanishing level spacing. Under this condition, one can make the replacement

$$\sum_{p \leq p_F} \to \frac{V}{(2\pi)^3} \int_{p \leq p_F} d^3 p \qquad (1.12)$$

in the right hand side of Eq. (1.11), to obtain

$$E = 2 \frac{V}{(2\pi)^3} 4\pi \int_0^{p_F} p^2 dp \frac{p^2}{2m_e} = N \frac{3}{5} \frac{p_F^2}{2m_e} . \qquad (1.13)$$

The corresponding energy density is

$$\epsilon = \frac{E}{V} = \frac{1}{(2\pi)^3} 4\pi \frac{p_F^5}{5m_e} . \qquad (1.14)$$

Note that from Eq. (1.10) it follows that the Fermi energy can be written in terms of the number density according to

$$\epsilon_F = \frac{p_F^2}{2m_e} = \frac{1}{2m_e} \left(3\pi^2 n_e\right)^{2/3} . \qquad (1.15)$$

The above equation can be used to define a density n_0 such that for $n_e \gg n_0$ the electron gas at given temperature T is fully degenerate. Full degeneracy is realised when the thermal energy T is much smaller than the Fermi energy ϵ_F, that is, when

$$n_e \gg n_0 = \frac{1}{3\pi^2} (2m_e\,T)^{3/2} \ . \tag{1.16}$$

The interior temperature of a typical star at the stage of hydrogen burning, such as the Sun, is $\sim 10^7$ K, and the corresponding value of n_0 is $\sim 10^{26}$ cm^{-3}; see Table 1.1. If we assume that the electrons originate from a fully ionised hydrogen gas, the *matter* density of the plasma comprising protons and electrons at number density n_0 is

$$\varrho = (m_p + m_e)\,n_0 \sim 200 \text{ g/cm}^3 \ , \tag{1.17}$$

m_p being the proton mass. This density is high for most stars in the early stage of hydrogen burning, while for ageing stars that have developed a substantial helium core the density

$$\varrho = (m_p + m_n + m_e)\,n_0 \sim 400 \text{ g/cm}^3 \ , \tag{1.18}$$

where $m_n \approx m_p$ denotes the neutron mass, can be largely exceeded within the core. Because white dwarfs have core densities of the order of 10^7 g/cm^3, in theoretical studies of their structure thermal energies can be safely neglected, the primary role being played by the degeneracy energy $p^2/2m_e$.

1.3.2   Pressure

The pressure P of the degenerate electron gas, that is, the force per unit area on the walls of the normalisation box, is defined in kinetic theory as the rate of momentum transferred by the electrons colliding on a surface of unit area.

Let us consider first one dimensional motion along the x-axis. The momentum transfer associated with the reflection of electrons carrying momentum p_x off the box wall during a time dt is

$$\frac{dp_x}{dt} = \frac{1}{2} \times (2p_x) \times (n_e \mathrm{v}_x L^2) \ , \tag{1.19}$$

v_x being the electron velocity. In the right-hand side of the above equation, the second factor is the momentum transfer associated with a single electron, while the third factor is the electron flux, that is, the number of electrons hitting the wall during the time dt. The factor $1/2$ accounts for the fact that half of the electrons go the *wrong* way, and do not collide with the box wall. The resulting pressure is

$$P(p_x) = \frac{1}{L^2} \frac{dp_x}{dt} = n_e p_x v_x = \frac{n_e p_x^2}{m_e} . \tag{1.20}$$

In the three-dimensional case the electron carries momentum $p = \sqrt{p_x^2 + p_y^2 + p_z^2}$, and we have to repeat the calculation carried out for the x-projection of the momentum transfer. Because the system is isotropic

$$\frac{p_x^2}{m_e} = p_x v_x = \frac{1}{3} (p_x v_x + p_y v_y + p_z v_z) = \frac{1}{3} (\mathbf{p} \cdot \mathbf{v}) = \frac{1}{3} \frac{p^2}{m_e} , \tag{1.21}$$

and Eq. (1.20) becomes

$$P(p) = \frac{1}{3} n_e(pv) = \frac{1}{3} \frac{N}{V} \frac{p^2}{m_e} . \tag{1.22}$$

The total pressure can now be obtained by averaging over all momenta. The result is

$$P = \frac{1}{N} 2 \sum_{p \leq p_F} P(p) = \frac{2}{3} \frac{1}{(2\pi)^3} 4\pi \int_0^{p_F} p^2 dp \, (pv) = \frac{p_F^5}{15\pi^2 m_e} . \tag{1.23}$$

Note that the above expression can also be obtained from the standard thermodynamic definition of pressure

$$P = - \left(\frac{\partial E}{\partial V} \right) , \tag{1.24}$$

using E given by Eq. (1.13) and $(\partial p_F / \partial V) = -p_F/(3V)$.

Equation (1.23) shows that, for any given density, the pressure of a degenerate Fermi gas decreases linearly as the mass of the constituent particle increases. For example, the pressure of an electron gas at number density n_e is ~2000 times larger than the pressure of a gas of nucleons, either neutrons or protons, at the same number density.

1.3.3 Relativistic Regime

So far, we have been assuming the electrons of the degenerate Fermi gas to be non relativistic. However, the properties of the system depend primarily on the distribution of quantum states, which is dictated by translation invariance, and turn out to be largely unaffected by this assumption. Releasing the non relativistic approximation simply amounts to substituting the non relativistic energy with its relativistic counterpart, that is, to perform the replacement

$$\frac{p^2}{2m_e} \rightarrow \sqrt{p^2 + m_e^2} - m_e . \tag{1.25}$$

The transition from the non relativistic regime to the relativistic one occurs when the electron energy becomes comparable to the electron rest mass, m_e. It is therefore possible to define a density n_c such that at $n_e \ll n_c$ the system is nonrelativistic, while $n_e \gg n_c$ corresponds to the relativistic regime.

The value of n_c can be readily found requiring that the Fermi energy at $n_e = n_c$ be equal to m_e. The resulting expression is

$$n_c = \frac{2^{3/2}}{3\pi^2} m_e^3 \sim 1.6 \times 10^{30} \text{ cm}^{-3} . \tag{1.26}$$

The energy density of a fully degenerate gas of relativistic electrons can be obtained from

$$\epsilon = 2 \frac{1}{(2\pi^3)} 4\pi \int_0^{p_F} p^2 dp \left[\sqrt{p^2 + m_e^2} - m_e \right] , \tag{1.27}$$

to be compared to Eqs. (1.13) and (1.14), while the corresponding expression of the pressure

$$P = \frac{2}{3} \frac{1}{(2\pi)^3} 4\pi \int_0^{p_F} p^2 dp \left(p \frac{\partial \epsilon_p}{\partial p} \right) , \tag{1.28}$$

can be derived from Eq. (1.23) using again $v = \partial \epsilon_p / \partial p$ with the relativistic ϵ_p.

Carrying out the integrations involved in Eqs. (1.27) and (1.28) one finds

$$\epsilon = \frac{\pi m_e}{\lambda_e^3} \left[t \left(2t^2 + 1 \right) \sqrt{t^2 + 1} - \ln \left(t + \sqrt{t^2 + 1} \right) - \frac{8t^3}{3} \right] , \tag{1.29}$$

and

$$P = \frac{\pi m_e}{\lambda_e^3} \left[\frac{1}{3} t \left(2t^2 - 3 \right) \sqrt{t^2 + 1} + \ln \left(t + \sqrt{t^2 + 1} \right) \right] , \tag{1.30}$$

where $\lambda_e = 2\pi / m_e$ is the electron Compton wavelength and the dimensionless quantity t is defined as

$$t = \frac{p_F}{m_e} = \frac{1}{m_e} \left(3\pi^2 n_e \right)^{1/3} , \tag{1.31}$$

as dictated by Eq. (1.15).

Equations (1.29) and (1.30) give the energy density and pressure of a fully degenerate electron gas as a function of the variable t, which can in turn be written in terms of the number density n_e using Eq. (1.31).

As a final remark, let us consider the possible relevance of electrostatic interactions. In a fully ionized plasma their effect can be estimated noting that the corresponding energy is

$$E_c = Z \frac{e^2}{\langle r \rangle} \propto Z \, e^2 \, n_e^{1/3} \, , \tag{1.32}$$

where Ze is the electric charge of the ions and $\langle r \rangle \propto n_e^{1/3}$ is the typical electron-ion separation distance. It follows that the ratio between E_c and the Fermi energy is given by

$$\frac{E_c}{\epsilon_F} \propto \frac{1}{n_e^{1/3}} \, . \tag{1.33}$$

The above equation shows that, for sufficiently high density, the contribution of electrostatic interactions becomes negligibly small. Because this condition is largely satisfied at the densities typical of white dwarfs, in these systems electrons can be safely described as a fully degenerate gas of non interacting electrons.

1.4   Significance of the Equation of State

The equation of state (EOS) is a nontrivial relation between the thermodynamic functions specifying the state of a physical system. The best known example of EOS is Boyle's *ideal gas law*, stating that, in the absence of interactions, the pressure of a collection of N point like classical particles enclosed in a volume V grows linearly with the temperature, T, and the average particle number density, $n = N/V$.

The ideal gas law provides a good description of very low-density systems. More generally, the EOS can be obtained expanding the pressure, P, in powers of density. The resulting expression reads

$$P = nT \left[1 + n B(T) + n^2 C(T) + \ldots \right] . \tag{1.34}$$

The coefficients appearing in the above series, known as *virial expansion*, are functions of temperature only. They describe the deviations from the ideal gas law and—at least in principle—should be derived from the underlying elementary interactions.

The EOS carries a great deal of dynamical information, and its knowledge allows to establish a link between measurable *macroscopic* quantities, such as pressure or temperature, and the forces acting between the constituents of the system at *microscopic* level.

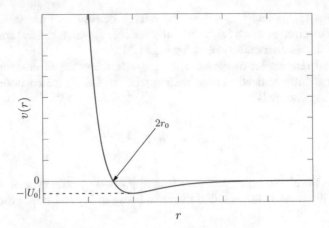

Fig. 1.1 Schematic representation of the Van der Waals potential

This dynamical content of the EOS can be clearly illustrated considering the van der Waals fluid, that is, a system consisting of particles interacting through a potential characterised by a strongly repulsive core and a weaker attractive tail; see Fig. 1.1. For $|U_0|/T \ll 1$, with U_0 being the strength of the attractive component of the potential, the EOS of the van der Waals fluid takes the simple form

$$P = \frac{nT}{1 - nb} - an^2 \, . \tag{1.35}$$

The right side of the above equation shows that the departure from Boyle's law is driven by the two parameters a and b, embodying the effects of interactions. Their values can be directly related to the potential $v(r)$ through [9]

$$a = \pi \int_{2r_0}^{\infty} |v(r)| \, r^2 dr \quad , \quad b = \frac{16}{3} \pi r_0^3 \, , \tag{1.36}$$

where $2r_0$ is the radius of the repulsive core; see Fig. 1.1. In this instance, the knowledge of the EOS allows one to determine the potential describing the dynamics at microscopic level.

In spite of its simplicity, the van der Waals EOS describes most features of both the gas and liquid phases of the system, as well as the nature of the phase transition.

1.5   Equation of State of White Dwarf Matter

Equation (1.30), which is applicable in both the non relativistic and relativistic regimes, is the EOS of the degenerate electron gas. It provides a link between the pressure, P, and the matter density, ϱ, which is in turn related to the electron number density n_e through

$$\varrho = \frac{m_p}{Y_e} n_e , \tag{1.37}$$

where Y_e is the number of electrons per nucleon in the system, that is, the ratio between the ion charge Z and its mass number A. For example, for a fully ionized helium plasma $Y_e = 2/4 = 0.5$, whereas for a plasma of iron, having $Z = 26$ protons and $A = 56$ nucleons, $Y_e = 26/56 = 0.464$.

The EOS of the fully degenerate electron gas takes a particularly simple form in the nonrelativistic limit, corresponding to $t \ll 1$, as well as in the extreme relativistic limit, corresponding to $t \gg 1$. From Eq. (1.30) we find

$$P = \frac{8}{15} \frac{\pi m_e}{\lambda_e^3} \left(\frac{3\pi^2 Y_e}{m_p} \right)^{5/3} \varrho^{5/3} . \tag{1.38}$$

for $e_F \ll m_e$ and

$$P = \frac{2}{3} \frac{\pi m_e}{\lambda_e^3} \left(\frac{3\pi^2 Y_e}{m_p} \right)^{4/3} \varrho^{4/3} . \tag{1.39}$$

for $e_F \gg m_e$.

An EOS that can be written in the form

$$P \propto \varrho^\Gamma , \tag{1.40}$$

as those of Eqs. (1.38) and (1.39), is said to be *polytropic*. The exponent Γ is referred to as *adiabatic index*, whereas the quantity n, defined through

$$\Gamma = 1 + \frac{1}{n} , \tag{1.41}$$

not to be confused with the particle number density, goes under the name of *polytropic index*.

The adiabatic index, whose extension to a generic EOS can be written in the form

$$\Gamma = \frac{d \ln P}{d \ln \varrho} , \tag{1.42}$$

is simply related to the compressibility of matter, χ, defined by the equation

$$\frac{1}{\chi} = -V \left(\frac{\partial P}{\partial V} \right) = \varrho \left(\frac{\partial P}{\partial \varrho} \right) . \tag{1.43}$$

through

$$\Gamma = \frac{1}{\chi P} \; . \tag{1.44}$$

From the above equations, it follows that the speed of sound in matter, c_s, can also be expressed in terms of χ, according to

$$c_s = \left(\frac{\partial P}{\partial \varrho} \right)^{1/2} = \frac{1}{\chi \varrho} \; . \tag{1.45}$$

The magnitude of the adiabatic index reflects an important property of the EOS, referred to as *stiffness*, which plays a critical role in determining many properties of compact stars. Larger stiffness corresponds to more incompressible matter.

1.6   Equilibrium of White Dwarfs and Chandrasekhar Limit

Le us consider a white dwarf consisting of a plasma of fully ionised helium at zero temperature. The pressure of the system, P, is provided mainly by the electrons, the contribution of the helium nuclei being negligible due to their large mass. For any given value of the matter density ϱ, P can be computed from Eqs. (1.30) and (1.31). Note that in this case $Y_e = 0.5$, implying $n_e = \varrho/2m_p$. The results of this calculation are shown by the diamonds in Fig. 1.2. For comparison, the non relativistic and extreme relativistic limits, given by Eqs. (1.38) and (1.39), are also

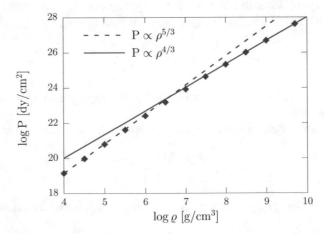

Fig. 1.2 Equation of state of a fully ionized helium plasma at zero temperature (diamonds). The dashed and solid line correspond to the non relativistic and extreme relativistic limits, respectively

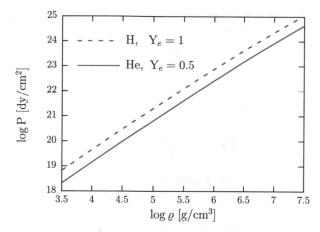

Fig. 1.3 Comparison between the EOS of a fully ionized plasma of helium (solid line) and hydrogen (dashed line)

shown by the solid and dashed line, respectively. Note that the value of matter density corresponding to n_c defined in Eq. (1.26) is $\varrho \sim 6.3 \times 10^6 \, \text{g/cm}^3$.

In order to show the sensitivity of the EOS to the value of Y_e, in Fig. 1.3 the EOS of the fully ionized helium plasma, having $Y_e = 0.5$ is compared to that of a hydrogen plasma, corresponding to $Y_e = 1$.

The surface gravity of white dwarfs—given by GM/R, with G being the gravitational constant—is small, of order $\sim 10^{-4}$. As a consequence, the structure of these systems can be studied assuming that they consist of a spherically symmetric fluid in hydrostatic equilibrium, neglecting relativistic effects altogether.

Consider a perfect fluid in thermodynamic equilibrium, subject only to gravity. Euler's equation can be written in the form

$$\frac{\partial \mathbf{v}}{\partial t} + (\mathbf{v} \cdot \nabla)\mathbf{v} = -\frac{1}{\varrho}\nabla P - \nabla \phi \, , \tag{1.46}$$

where ϱ is the fluid density and ϕ is the gravitational potential, satisfying Poisson's equation

$$\nabla^2 \phi = 4\pi G \varrho \, . \tag{1.47}$$

Equation (1.46) describes the motion of a fluid in which processes leading to energy dissipation due to viscosity—that is, internal friction—and heat exchange between different regions can be neglected. In the rest frame of the fluid, in which $\mathbf{v} = 0$, Eluer's equation reduces to

$$\nabla P = -\varrho \, (\nabla \phi) \, . \tag{1.48}$$

In the case of a spherically symmetric fluid, Eqs. (1.47) and (1.48) can be written in the form

$$\frac{dP}{dr} = -\varrho \frac{d\phi}{dr} ,\tag{1.49}$$

and

$$\frac{1}{r^2} \frac{d}{dr} \left(r^2 \frac{d\phi}{dr} \right) = 4\pi G \varrho .\tag{1.50}$$

Substitution of Eq. (1.49) in Eq. (1.50) yields

$$\frac{1}{r^2} \frac{d}{dr} \left(\frac{r^2}{\varrho} \frac{dP}{dr} \right) = -4\pi G \varrho ,\tag{1.51}$$

implying

$$\frac{dP}{dr} = -\varrho(r) \frac{GM(r)}{r^2} ,\tag{1.52}$$

with $M(r)$ given by

$$M(r) = 4\pi \int_0^r \varrho(r') r'^2 dr' .\tag{1.53}$$

The above result simply states that, at equilibrium, the gravitational force acting on a volume element at distance r from the center of the star is balanced by the force produced by the spacial variation of the pressure.

Given an EOS $P = P(\varrho)$, Eq. (1.52) can be integrated numerically for any value of the central density, $\varrho(0) = \varrho_c$, to obtain the radius of the star, R, defined as the value of r corresponding to vanishingly small pressure. The mass of the star can then be obtained from Eq. (1.53) setting $r = R$.

For polytropic EOS Eqs. (1.52) and (1.53) reduce to the Lane-Emden equation [10], whose integration with the polytropic index $n = 3/2$—corresponding to the non relativistic regime—yields the relation

$$M = \frac{2.79}{Y_e^2} \left(\frac{\varrho_c}{\overline{\varrho}} \right)^{1/2} M_\odot ,\tag{1.54}$$

where $\overline{\varrho}$ denotes the matter density corresponding to the electron number density of Eq. (1.26). The resulting values of M agree with the results of astronomical observations of white dwarfs. However, it is very important to realise that the Lane-Emden equation predicts the existence of equilibrium configurations for any values of the star mass.

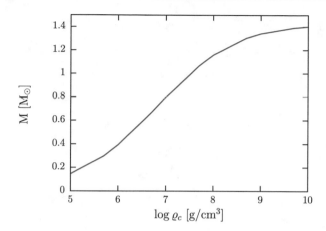

Fig. 1.4 Dependence of the mass of a white dwarf upon its central density, obtained from the integration of Eq. (1.52) using the equation of state of a fully ionized helium plasma

In 1931 Chandrasekhar pointed out that, due to the large Fermi energies, the non relativistic treatment of the electron gas was not justified [8]. Replacing the EOS of Eq. (1.38) with its relativistic counterpart, Eq. (1.39), he predicted the existence of a maximum mass for white dwarfs. If the mass exceeds this limiting value, gravitational attraction prevails on the pressure gradient, and the star becomes unstable against gravitational collapse.

The dependence of the mass of a white dwarf on its central density, obtained from integration of Eq. (1.52) using the equation of state of a fully ionized helium plasma, is illustrated in Fig. 1.4. The figure shows that the mass increases as the central density increases, until a value $M \sim 1.44 \, M_\odot$ is reached at $\varrho_{Ch} \sim 10^{10} \, g/cm^3$. This value is close to the one found by Chandrasekhar.

It has to be kept in mind, however, that at $\varrho \sim 10^8 \, g/cm^3$ the *neutronization* process—to be discussed at a later stage—sets in, and the validity of the description in terms of a helium plasma breaks down. At $\varrho \geq 10^8 \, g/cm^3$, matter does not support pressure as effectively as predicted by the equation of state of the helium plasma. As a consequence, a more realistic estimate of the limiting mass, generally referred to as the *Chandrasekhar mass*, is given by the mass corresponding to a central density of $10^8 \, g/cm^3$, that turns out to be $\sim 1.2 \, M_\odot$.

Part II
Neutron Stars

Neutron Star Structure

2

Abstract

The existence of compact stars, having density similar to that of atomic nuclei, was argued even before the discovery of the neutron in 1932. While sharing some features of the white dwarfs, these stars—which were eventually observed as radio pulsars in 1967—feature a much stronger surface gravity. Their equilibrium, described by Einstein's equations of general relativity, is determined by a pressure other that the degeneracy pressure of a Fermi gas of neutrons. Because the neutron star density spans many order of magnitudes, the ground state of matter in the star interior corresponds to a variety of different configurations, involving different constituents. From a Coulomb lattice of nuclei immersed in a electron gas, to a mixed phase featuring in addition a neutron gas, to a uniform fluid consisting mainly of neutrons, with a small fraction of protons and leptons. Finally, in the inner region—the density of which largely exceeds nuclear density—new phases of matter, non observable in standard terrestrial conditions, are expected to appear.

2.1 Discovery of Neutron Stars

The temperatures attained in stars with initial mass larger than $\sim 8\ M_\odot$ are high enough to bring nucleosynthesis to its final stage, featuring the formation of a core of ^{56}Fe; see Table 1.1. If the mass of the core exceeds the Chandrasekhar mass, the pressure of the degenerate electron gas is no longer sufficient to balance the gravitational attraction, and the star evolves towards the formation of a neutron star or a black hole.

© The Author(s), under exclusive license to Springer Nature Switzerland AG 2023
O. Benhar, *Structure and Dynamics of Compact Stars*, Lecture Notes
in Physics 1019, https://doi.org/10.1007/978-3-031-35628-5_2

The formation of the core in massive stars is accompanied by the appearance of neutrinos, produced in the weak decay process

$$^{56}\text{Ni} \rightarrow {}^{56}\text{Fe} + 2e^+ + 2\nu_e \, . \tag{2.1}$$

Neutrinos do not have appreciable interactions with the surrounding matter and leave the core region carrying away energy. As a consequence, neutrino emission contributes to the collapse of the core. Other processes leading to a decrease of the pressure are electron capture

$$p + e^- \rightarrow n + \nu_e \, , \tag{2.2}$$

the main effect of which is the disappearance of relativistic electrons carrying large kinetic energies, and iron photodisintegration

$$\gamma + {}^{56}\text{Fe} \rightarrow 13\,{}^4\text{He} + 4\,n \, , \tag{2.3}$$

which is an endothermic reaction.

Due to the combined effect of the above mechanisms, when the mass exceeds the Chandrasekhar limit the core collapses within fractions of a second to reach densities comparable with the central density of atomic nuclei, $\varrho_0 = 2.6 \times 10^{14}$ g/cm^3.

At this stage the core behaves as a giant nucleus, made mostly of neutrons, and reacts elastically to further compression producing a strong shock wave, which throws away a significant fraction of matter in the outer layers of the star. Nucleosynthesis of elements heavier than Iron is believed to take place during this explosive phase.

The sequence of events described above leads to the appearance of a *supernova*. The luminosity of a typical supernova grows very fast for ~20 days, until it reaches a value exceeding the luminosity of the sun by a factor $\sim 10^9$, and then decreases by a factor $\sim 10^2$ within few months. The final result of the explosion is the formation of a *nebula*, the center of which is occupied by the remnant of the core, that is, a neutron star.

The existence of compact astrophysical objects consisting mainly of neutrons— reportedly argued by Bohr, Landau and Rosenfeld shortly after the discovery of the neutral constituent of the atomic nucleus [11]—was first discussed by Landau in 1932 [12, 13]. In 1934, Baade and Zwicky suggested that a neutron star may be formed in the aftermath of a supernova explosion [14, 15]. Finally, in 1968 the newly observed pulsars, radio sources blinking on and off at a constant frequency, were identified with rapidly rotating neutron stars [16].

The results of pioneering studies carried out in 1939 by Tolman, and, independently, by Oppenheimer and Volkoff [17, 18] demonstrated that within the framework of general relativity the mass of a star consisting of noninteracting neutrons cannot exceed ~0.8 $M_\odot$. The observation that this maximum mass, analog to the Chandrasekhar mass of white dwarfs, turns out to be much smaller that the observed neutron star masses—typically ~ 1.4 $M_\odot$—clearly indicates that in neutron stars the stability against gravitational collapse requires a pressure other than the degeneracy pressure of a neutron gas. The origin of this pressure has to be ascribed to hadronic interactions.

Unfortunately, the need of including the effects of microscopic dynamics in the EOS is confronted with the daunting complexity of the fundamental theory of strong interactions. As a consequence, all available description of the EOS of neutron star matter are obtained within models, based on the theoretical knowledge of the underlying dynamics and constrained, as much as possible, by empirical data.

2.2   Overview of Neutron Star Composition

The internal structure of a neutron star, schematically illustrated in Fig. 2.1, is believed to feature a sequence of layers of different density and composition.

The properties of matter in the outer crust, corresponding to densities ranging from $\sim 10^7$ g/cm^3 to the *neutron drip* density $\varrho_d = 4 \times 10^{11}$ g/cm^3, can be largely inferred from nuclear data. On the other hand, models of the EOS in the inner crust, spanning the region $4 \times 10^{11} < \varrho < 2 \times 10^{14}$ g/cm^3, are necessarily based on extrapolations of the available empirical information, since the extremely neutron rich nuclei appearing in this density regime are not observed in ordinary Earth conditions.

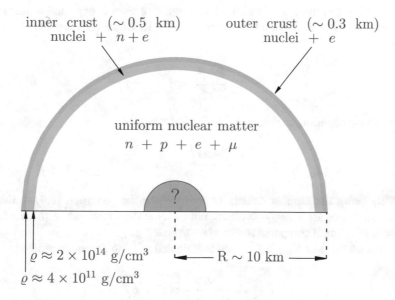

Fig. 2.1 Schematic representation of the neutron star cross section. For reference, note that the central density of atomic nuclei is $\sim 2 \times 10^{14}$ g/cm^3

The density of the neutron star core ranges between the nuclear density ϱ_0, at the boundary with the inner crust, and a central value that can be as large as $1-4 \times 10^{15}$ g/cm^3. All models of EOS based on hadronic degrees of freedom predict that in the density range $\varrho_0 \lesssim \varrho \lesssim 2\varrho_0$ neutron star matter consists mainly of neutrons, with the admixture of a small number of protons, electrons and muons. For any given nucleon density, the fraction of protons and leptons is determined by the requirements of equilibrium with respect to neutron β-decay and charge neutrality.

The picture may change significantly at larger density, with the appearance of strange baryons produced in weak interaction processes. For example, although the mass of the Σ^- exceeds the neutron mass by more than 250 MeV, the reaction $n + e^- \rightarrow \Sigma^- + \nu_e$ is energetically allowed as long as the sum of the neutron and electron chemical potentials becomes equal to the Σ^- chemical potential.

Finally, as nucleons are known to be composite objects of size $\sim 0.5 - 1.0$ fm, corresponding to an interior density $\sim 10^{15}$ g/cm^3, it is expected that if the density in the neutron star core reaches this value matter undergoes a transition to a new phase, in which quarks are no longer clustered into nucleons.

The theoretical description of matter in the outer and inner neutron star crust will be outlined in the following sections, whereas the region corresponding to supranuclear density will be described in Chap. 3.

2.2.1 Outer Crust

In a fully ionised plasma of nuclei with charge Z, a solid is expected to form when the ratio between the Coulomb and thermal energies of the ions becomes large, that is, when

$$\frac{Z^2 e^2}{T r_L} \gg 1 . \tag{2.4}$$

In the above equation, the radius r_L is defined through

$$n_I \frac{4\pi r_L^3}{3} = 1 , \tag{2.5}$$

with n_I being the number density of ions. When the condition (2.4) is fulfilled, Coulomb forces are weakly screened and become dominant, while the fluctuation of the ions is small compared to average ion spacing r_L.

Equation (2.4) implies that a solid is expected to form at temperature

$$T < T_m = \frac{Z^2 e^2}{r_L} \propto Z^2 e^2 n_I^{1/3} . \tag{2.6}$$

For example, in the case of ^{56}Fe at density $\sim 10^7$ g/cm^3 solidification occurs at temperatures below 10^8 K, with the Coulomb energy being minimised by a Body

Centered Cubic, or BCC, lattice. As the density increases, r_L decreases, so that the condition for solidification continues to be satisfied. If matter density reaches the region 10^7–10^{11} g/cm^3, however, the large kinetic energy of the relativistic electrons shifts the energy balance favouring inverse β-decay, that is, electron capture. This process is called *neutronisation*, because it leads to the appearance of nuclides with increasing neutron excess $N = (A - Z)/Z$. For example, in the sequence

$$\text{Fe} \rightarrow \text{Ni} \rightarrow \text{Se} \rightarrow \text{Ge} \;. \tag{2.7}$$

the value of N increases from 1.15 in Iron to 1.56 in Germanium.

Before proceeding to the analysis of the neutronisation process in the neutron star crust, let us discuss a simple but instructive example, that will allows to introduce some concepts and procedures to be used in the following sections.

Inverse β-decay

Consider a gas of noninteracting protons and electrons at $T = 0$. Neutronisation results from the occurrence of electron capture processes, that is, inverse β-decay turning protons into neutrons

$$p + e^- \rightarrow n + \nu_e \;. \tag{2.8}$$

Assuming the neutrinos to be massless and non interacting, the above process is energetically favoured as soon as the electron energy equals the neutron-proton mass difference

$$\Delta m = m_n - m_p = 939.565 - 938.272 = 1.293 \text{ MeV} \;. \tag{2.9}$$

As a consequence, the value of n_e corresponding to the onset of inverse β-decay can be estimated from

$$\sqrt{p_{F_e}^2 + m_e^2} = \Delta m \;, \tag{2.10}$$

with

$$p_{F_e} = (3\pi^2 n_e)^{1/3} \;, \tag{2.11}$$

leading to

$$n_e = \frac{1}{3\pi^2} \left(\Delta m^2 - m_e^2 \right)^{3/2} \approx 7 \times 10^{30} \text{cm}^{-3} \;. \tag{2.12}$$

The corresponding mass density for a system having $Y_e \sim 0.5$, is $\varrho \approx 2.4 \times 10^7$ g/cm^3.

Let us now determine the ground state of a charge-neutral system consisting of protons, electrons and neutrons in equilibrium with respect to process (2.8), which conserves both baryon number density n_B and electric charge. All interactions other than weak interactions will be neglected.

For any given value of n_B, the ground state is found by minimising the total energy density of the systems, $\epsilon(n_p, n_n, n_e)$, n_p and n_n being the proton and neutron number density, with the constraints $n_B = n_p + n_n$ and $n_p = n_e$ following from conservation of baryon number and charge neutrality, respectively.

Let us define the function

$$F(n_p, n_n, n_e) = \epsilon(n_p, n_n, n_e) + \lambda_B \left(n_B - n_p - n_n\right) + \lambda_Q \left(n_p - n_e\right) , \qquad (2.13)$$

where ϵ is the total energy density of the system, while λ_B and λ_Q are Lagrange multipliers.

The minimum of F corresponds to the values of n_p, n_n and n_e satisfying the conditions

$$\frac{\partial F}{\partial n_p} = \frac{\partial F}{\partial n_n} = 0 \ , \qquad \frac{\partial F}{\partial n_e} = 0 \qquad (2.14)$$

as well as the additional constraints

$$\frac{\partial F}{\partial \lambda_B} = \frac{\partial F}{\partial \lambda_Q} = 0 . \qquad (2.15)$$

From the definition of the chemical potential of particles of species i

$$\mu_i = \left(\frac{\partial E}{\partial N_i}\right)_{V, n_{j \neq i}} = \left(\frac{\partial \epsilon}{\partial n_i}\right)_{V, n_{j \neq i}} , \qquad (2.16)$$

with $i, j = p, n, e$, it follows that Eq. (2.14) entail the relations

$$\mu_p - \lambda_B + \lambda_Q = 0 \ , \quad \mu_n - \lambda_B = 0 \ , \quad \mu_e - \lambda_Q = 0 , \qquad (2.17)$$

which in turn lead to the condition of chemical equilibrium

$$\mu_n - \mu_p = \mu_e . \qquad (2.18)$$

In the case of noninteracting particles at $T = 0$, the above chemical potentials take the simple form

$$\mu_i = \frac{\partial \epsilon}{\partial n_i} = \frac{2}{(2\pi)^3} \left(\frac{\partial p_{F_i}}{\partial n_i}\right) \frac{\partial}{\partial p_{F_i}} 4\pi \int_0^{p_{F_i}} p^2 \, dp \, \sqrt{p^2 + m_i^2}$$

$$= \frac{8\pi}{(2\pi)^3} \left(\frac{\partial n_i}{\partial p_{F_i}}\right)^{-1} p_{F_i}^2 \sqrt{p_{F_i}^2 + m_i^2} = \sqrt{p_{F_i}^2 + m_i^2} , \qquad (2.19)$$

with $i = n$, p, and e. Defining now the proton, electron and neutron fractions specifying the composition of the system as

$$x_p = x_e = \frac{n_p}{n_B} = \frac{n_p}{n_p + n_n} \quad , \quad x_n = \frac{n_n}{n_B} = 1 - x_p , \tag{2.20}$$

one can rewrite

$$p_{F_p} = p_{F_e} = \left(3\pi^2 x_p n_B\right)^{1/3} \quad , \quad p_{F_n} = \left[3\pi^2 \left(1 - x_p\right) n_B\right]^{1/3} . \tag{2.21}$$

For fixed baryon density, use of the above definitions in Eq. (2.19) and substitution of the resulting chemical potentials into Eq. (2.18) leads to an equation in the single unknown variable x_p. Hence, for any given n_B the requirements of chemical equilibrium and charge neutrality uniquely determine the fraction of protons in the system.

Once the value of x_p is known, the neutron, proton and electron number densities can be readily evaluated, and the pressure

$$P = P_n + P_p + P_e \tag{2.22}$$

is obtained from Eq. (1.28).

Figure 2.2 shows the proton and neutron number densities, n_p and n_n displayed as a function of the matter density ϱ in log–log scale. Recall that, owing to charge neutrality, $n_e = n_p$. It can be seen that in the range $10^5 \le \varrho \le 10^7$ g/cm³ there are protons only and $\log n_p$ grows linearly with $\log \varrho$. At $\varrho \sim 10^7$ g/cm³ neutronisation

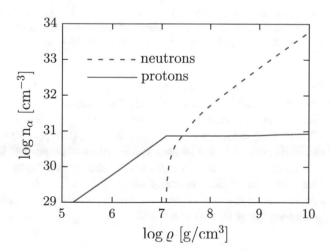

Fig. 2.2 Number density of noninteracting protons and neutrons in β-equilibrium as a function of matter density

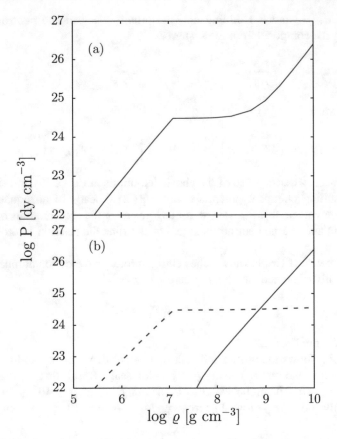

Fig. 2.3 (a) Equation of state of a charge-neutral mixture of noninteracting neutrons, electrons and protons in β-equilibrium. (b) Density dependence of the neutron (solid line) and electron (dashed line) contributions to the pressure of charge-neutral β-stable matter

sets in, and the neutron number density begins to steeply increase. At $\varrho > 10^7$ g/cm^3 n_p stays nearly constant, while neutrons become dominant at $\varrho > 10^9$ g/cm^3.

The equation of state of charge-neutral β-stable matter is shown in the upper panel of Fig. 2.3. It is apparent that the pressure, shown in panel (a), remains nearly constant as matter density increases by almost two orders of magnitude in the range $10^7 \leq \varrho \leq 10^9$ g/cm^3. This feature is explained by the density dependence of the electron and neutron contributions to the pressure, shown in panel (b). Note that, because charge neutrality requires $n_p = n_e$, the proton pressure is smaller than the electron pressure by a factor $(m_p/m_e) \sim 2000$.

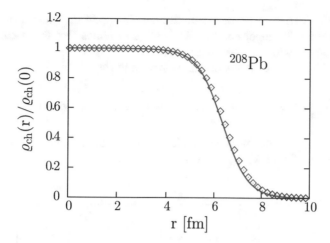

Fig. 2.4 Nuclear charge distribution of the nucleus ^{208}Pb normalized to $Z/\varrho(0)$, with $Z = 82$. The solid line has been obtained using the parametrisation of Eq. (2.24), while the diamonds represent the results of a model independent analysis of electron scattering data

Neutronisation

The description of charge-neutral β-stable matter in terms of a mixture of degenerate Fermi gases of neutrons, protons and electrons is strongly oversimplified. In reality, electron capture changes a nucleus with given charge Z and mass number A into a different nucleus with the same A and charge $(Z-1)$. Moreover, if the new nucleus is metastable, two-step processes such as

$$^{56}_{26}\text{Fe} \rightarrow {}^{56}_{25}\text{Mn} \rightarrow {}^{56}_{24}\text{Cr} \tag{2.23}$$

can take place. In this case, chemical equilibrium is driven by the mass difference between neighbouring nuclei, rather than the difference between the neutron and proton chemical potentials.

The charge density distributions and binding energies inferred from nuclear data, $\varrho_{ch}(r)$ and $M(Z, A)$, exhibit two very important features

- The charge density is nearly constant within the nuclear volume—its value being roughly the same for all stable nuclei—and drops from ~90% to ~10% of the maximum over a distance $R_T \sim 2.5\,\text{fm}$,[1] independent of A, called surface thickness; see Fig. 2.4. It can be accurately parametrised in the form

$$\varrho_{ch}(r) = \varrho_0 \, \frac{1}{1 + e^{(r-R)/D}}\,, \tag{2.24}$$

---

[1] Recall: 1 fm = 10×10^{-13} cm.

where $R = r_0 A^{1/3}$, with $r_0 = 1.07$ fm, and D $= 0.54$ fm. Note that the nuclear charge radius is proportional to $A^{1/3}$, which implies that the nuclear volume grows linearly with the mass number A.

- The positive binding energy per nucleon, defined as

$$\frac{B(Z, A)}{A} = \frac{1}{A} \left[Z m_p + (A - Z) m_n + Z m_e - M(Z, A) \right] , \qquad (2.25)$$

where $M(Z, A)$ is the measured nuclear mass, is almost constant for $A \geq 12$, its value being ~ 8.5 MeV; see Fig. 2.5.

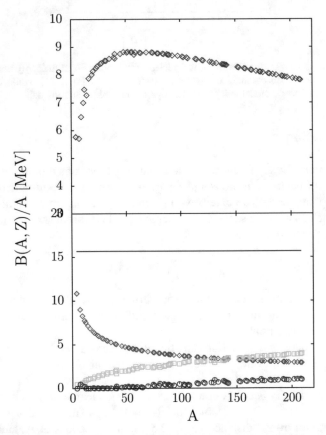

Fig. 2.5 Upper panel: A-dependence of the binding energy per nucleon of stable nuclei, evaluated according to Eq. (2.26) with $a_V = 15.67$ MeV, $a_s = 17.23$ MeV, $a_c = .714$ MeV, $a_A = 93.15$ MeV and $a_p = 11.2$ MeV. Lower panel: the solid horizontal line shows the magnitude of the volume contribution to the binding energy per nucleon, whereas the A-dependence of the surface, Coulomb and symmetry terms are represented by diamonds, squares and circles, respectively

The A and Z dependence of $B(Z, A)$ can be accurately parametrised by the von Weitzäcker *semi empirical mass formula* [19], whose derivation is based on the liquid drop model of the nucleus [20]

$$\frac{B(Z, A)}{A} = \frac{1}{A}\left[a_V A - a_S A^{2/3} - a_C \frac{Z^2}{A^{1/3}} - a_A \frac{(A - 2Z)^2}{4A} + \lambda \, a_P \frac{1}{A^{1/2}}\right].$$

(2.26)

The first term in square brackets, proportional to A, is called *volume term* and describes the bulk energy of nuclear matter. The second term, proportional to the nuclear radius squared, is associated with the surface energy, while the third one accounts for the Coulomb repulsion between Z protons uniformly distributed within a sphere of radius R. The fourth term, that goes under the name of *symmetry energy* is required to describe the experimental observation that stable nuclei tend to have the same number of neutrons and protons. In addition, even-even nuclei, that is, nuclei having even Z and even $N = A - Z$, tend to be more stable than even-odd or odd-odd nuclei. This property is accounted for by the last term in the right-hand side of Eq. (2.26), with $\lambda = -1$, 0 and +1 for even-even, even-odd and odd-odd nuclei, respectively. Figure 2.5 illustrates the different contributions to $B(Z, A)/A$, evaluated using Eq. (2.26).

The semiempirical nuclear mass formula of Eq. (2.26) can be used to obtain a qualitative description of the neutronisation process. The total energy density of the system consisting of nuclei of mass number A and charge Z arranged in a lattice and surrounded by a degenerate electron gas is given by

$$\epsilon_T(n_B, A, Z) = \epsilon_e + \left(\frac{n_B}{A}\right)[M(Z, A) + \epsilon_L],$$

(2.27)

where ϵ_e is the energy density of the electron gas, Eq. (1.27), n_B and (n_B/A) denote the number densities of nucleons and nuclei, respectively, and ϵ_L is the electrostatic lattice energy per site. As a first approximation, the contribution of ϵ_L will be neglected.

At any given nucleon density n_B, the equilibrium configuration corresponds to the values of A and Z that minimise $\epsilon_T(n_B, A, Z)$, that is, to A and Z such that

$$\left(\frac{\partial \epsilon_T}{\partial Z}\right)_{n_B} = 0, \quad \left(\frac{\partial \epsilon_T}{\partial A}\right)_{n_B} = 0.$$

(2.28)

Combining the above relationships and using Eq. (2.26) one finds

$$Z \approx 3.54 \, A^{1/2}.$$

(2.29)

Once Z is known as a function of A, any of the two relationships (2.28) can be used to obtain the mass number as a function of baryon number density n_B.

The mass number A turns out to be an increasing function of n_B, implying that Z also increases with n_B, but at a slower rate. Hence, nuclei become more massive and more and more neutron rich as the nucleon density increases.

Obviously, the above discussion is still somewhat simplified. In reality, A and Z are *not* continuous variables and the total energy has to be minimised using the measured nuclear masses, rather than the parametrisation of Eq. (2.26). Furthermore, the lattice energy, that can be written in the form

$$\epsilon_L = -K_L \frac{(Ze)^2}{r_s} , \tag{2.30}$$

must be also taken into account. Here, r_s is related to the number density of nuclei through $4\pi r_s^3/3 = (n_B/A)^{-1}$, while the value of the constant K_L depends on lattice topology.

The total energy density of a BCC lattice, corresponding to the lowest energy, can be written as a function of baryon number density in the form

$$\epsilon_T(n_B, A, Z) = \epsilon_e + \left(\frac{n_B}{A}\right)\left[M(Z, A) - 1.4442(Ze)^2 \left(\frac{n_B}{A}\right)^{1/3}\right], \tag{2.31}$$

where, for matter density exceeding $\sim 10^6$ g/cm^3, the extreme relativistic limit of the energy density of an electron gas at number density $n_e = Zn_B/A$ derived in Sect. 1.3

$$\epsilon_e = \frac{3}{4}\left(Z\frac{n_B}{A}\right)^{4/3} , \tag{2.32}$$

must be used.

Collecting together the results of Eqs. (2.30)–(2.32) and expressing n_B in units of $n_{B_0} = 10^{-9}$ fm^{-3}—the number density corresponding to a matter density $\sim 10^6$ g/cm^3—the total *energy per nucleon*, ϵ_T/n_B, can be rewritten in units of MeV as

$$\frac{\epsilon_T}{n_B} = \frac{M(Z, A)}{A} + \frac{1}{A^{4/3}}\left[0.4578\,Z^{4/3} - \frac{Z^2}{480.74}\right]\left(\frac{n_B}{n_{B_0}}\right)^{1/3} . \tag{2.33}$$

Because the average energy per nucleon in a nucleus is about 930 MeV, the nuclear mass can be conveniently written in units of MeV in the form $M(Z, A)/A = 930 + \Delta$. For nuclides that are not very far from stability, the values of Δ are available in form of tables based on actual measurements or extrapolations of the experimental data.

In practice, ϵ_T/n_B of Eq. (2.33) is computed for a given nucleus, that is, for given A and Z, as a function of n_B, and plotted versus $1/n_B$; see Fig. 2.6. The curves corresponding to different nuclei are then compared and the nucleus corresponding to the minimum energy at given n_B can be readily identified. For example, the curves of Fig. 2.6 show the behavior of the energy per particle corresponding to ^{62}Ni and

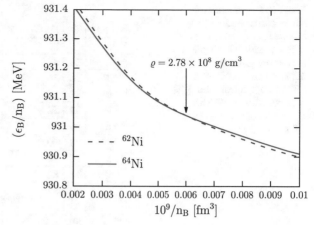

Fig. 2.6 Total energy per nucleon of a BCC lattice of ^{62}Ni (dashed line) and ^{64}Ni (solid line) nuclei surrounded by an electron gas, evaluated using Eq. (2.33) and plotted versus the inverse nucleon number density

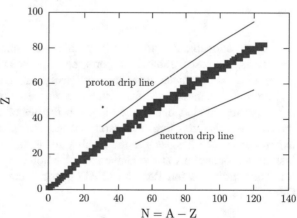

Fig. 2.7 Chart of the nuclides. The filled squares represent the stable nuclei as a function of charge, Z, and number of neutrons, $N = A - Z$. The solid lines correspond to the estimated proton and neutron drip lines

^{64}Ni, having $A - Z = 34$ and 36, respectively. It is apparent that a first order phase transition is taking place around the point where the two curves cross one another. The densities at which the phase transition appears and terminates can be precisely determined using Maxwell's double tangent construction. This method essentially amounts to drawing a straight line tangent to the convex curves corresponding to the two nuclides. In a first order transition the pressure remains constant as the density increases. Hence, as all points belonging to the tangent of Maxwell's construction correspond to the same pressure, and the appearance and termination of the phase transition simply correspond to the points of tangency. Note that, as expected, at higher density the nucleus with the largest number of neutrons yields the lower energy.

It has to be pointed out that there are limitations to the approach described above. Some of the nuclides involved in the minimisation procedure have ratios Z/A very different from those corresponding to stable nuclei, whose typical value is $\sim$0.5; see Fig. 2.7. As a consequence, in these cases the accuracy of the extrapolated masses

Table 2.1 Sequence of
nuclei corresponding to the
ground state of matter as a
function of density. Nuclear
masses are given by
$M(Z, A)/A = (930 + \Delta)$
MeV. Adapted from Ref. [21]
with permissions, © The
University of Chicago 1971.
All rights reserved

Nuclide	Z	$N = A - Z$	Z/A	Δ (MeV)	ϱ_{max} (g/cm^{-3})
^{56}Fe	26	30	0.4643	0.1616	8.1×10^6
^{62}Ni	28	34	0.4516	0.1738	2.7×10^8
^{64}Ni	28	36	0.4375	0.2091	1.2×10^9
^{84}Se	34	50	0.4048	0.3494	8.2×10^9
^{82}Ge	32	50	0.3902	0.4515	2.1×10^{10}
^{84}Zn	30	54	0.3750	0.6232	4.8×10^{10}
^{78}Ni	28	50	0.3590	0.8011	1.6×10^{11}
^{76}Fe	26	50	0.3421	1.1135	1.8×10^{11}
^{124}Mo	42	82	0.3387	1.2569	1.9×10^{11}
^{122}Zr	40	82	0.3279	1.4581	2.7×10^{11}
^{120}Sr	38	82	0.3166	1.6909	3.7×10^{11}
^{118}Kr	36	82	0.3051	1.9579	4.3×10^{11}

may be questionable. Obviously, this problem becomes more and more critical as
the density increases. Table 2.1 reports the sequence of nuclides corresponding to
the ground state of matter at subnuclear density, as a function of nucleon density.

The study of nuclei far from stability, carried out using radioactive nuclear
beams, is regarded as one of the highest priorities of nuclear physics research. In
the coming years, this line of research will be extensively pursued using dedicated
experimental facilities, such as the Facility for Rare Isotope Beams (FRIB), which
just started operations at the University of Michigan, USA [22], and the Facility
for Antiproton and Ion Research (FAIR), under construction at GSI Darmstadt,
Germany [23].

2.2.2 Inner Crust

Table 2.1 shows that, as the density increases, the nuclides corresponding
to the ground state of matter become more and more neutron rich. At
$\varrho \sim 4.3 \times 10^{11}$ g/cm^3 the lowest energy configuration consists of a Coulomb
lattice of ^{118}Kr nuclei—having proton to neutron ratio $Z/N \approx 0.44$ and a slightly
negative neutron chemical potential, coinciding with the neutron Fermi energy—
surrounded by a degenerate electron gas with chemical potential $\mu_e \approx 26$ MeV.

At larger densities a new regime sets in, because there are no negative energy
levels left to accommodate the neutrons created by electron capture. Neutrons begin
to occupy positive energy levels, and are no longer bound. As a consequence, they
begin to *drip* out of the nuclei, filling the space between them. At these densities the
ground state corresponds to a mixture of two phases: matter consisting of neutron
rich nuclei at density ϱ_{nuc}, or phase I, and a neutron gas at density ϱ_{NG}, or phase II.

The equilibrium conditions are

$$(\mu_n)_I = (\mu_n)_{II} = \mu_n \tag{2.34}$$

and

$$\mu_p = \mu_n - \mu_e , \tag{2.35}$$

where $(\mu_n)_I$ and $(\mu_n)_{II}$ denote the neutron chemical potential in the neutron gas and in the matter of nuclei, respectively.

The structure of the ground state in the neutron drip regime is determined by the densities ϱ, ϱ_{nuc} and ϱ_{NG}, the proton to neutron ratio of nuclei in phase I, and the geometric arrangement of matter in phases I and II. In the 1980s, Ravenhall et al. [24] argued that the transition from the outer crust the core region is smooth, and associated with the appearance of a mixed phase dubbed *nuclear pasta*. More recent studies have confirmed this picture, suggesting that at densities $4.3 \times 10^{10} \lesssim \varrho \lesssim 0.75 \times 10^{14}$ g/cm^3 the matter in phase I is arranged in spheres immersed in the electron and neutron gases, whereas at $0.75 \times 10^{14} \lesssim \varrho \lesssim 1.2 \times 10^{14}$ g/cm^3 the balance between Coulomb and surface energies favours the emergence of more complex configurations, featuring rods of matter in phase I or alternating layers of matter in phase I and phase II. At larger densities the role of phases I and II is reversed, and, finally, at $\varrho \gtrsim 1.2 \times 10^{14}$ g/cm^3 the two phases are no longer separated. In this regime the ground state of matter is a homogeneous fluid of neutrons, protons and electrons; see Fig. 2.8. The properties of the *pasta* phase will be discussed further in the context of the transition from nuclear matter to quark matter; see Chap. 4.

Fig. 2.8 Schematic representation of the nuclear pasta phase in the neutron star inner crust. Panels (**a**)–(**c**) represent droplets of matter in phase I, cylindrical rods of matter in phase I, and alternate slabs of matter in phases I and II, respectively. Panels (**d**) and (**e**) illustrate the structures appearing in the density region in which the role of the two phases is reversed. Reprinted from Ref. [25] with permissions, ©APS 2013. All rights reserved

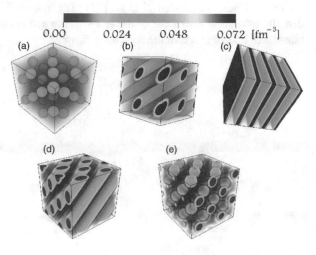

Superfluidity and Superconductivity

The interaction between nucleons, to be discussed in detail in Chap. 3, results from the combination of a very strong repulsive force at short range and a weaker attractive force at longer distance. At the relatively low density of the neutron phase in the neutron star inner crust, however, the interparticle separation is much larger than the range of the repulsive interactions, and attractive forces dominate. In this regime, both conditions for the occurrence of superfluidity in fermionic systems—that is, strong degeneracy and the presence of an attractive interaction leading to the formation of Cooper pairs—are expected to be fulfilled; see Refs. [26–28]. The results of theoretical calculations indicate that pairing of neutrons in the $^1S_0$ state[2] does, in fact, occur in the inner crust at densities $\varrho < 0.016$ g/cm^{-3}; see, e.g., Refs. [29, 30].

The onset of the superfluid phase does not have a significant impact on the EOS determining the neutron star equilibrium properties. The condensation energy—that is, the difference between the energies of the normal and superfluid states—turns out to be small, although not totally negligible, with respect to the typical energies of the normal phase. The main effect of the superfluid transition is the appearance of a gap in the energy spectrum located at the Fermi surface, which leads to a reduction of the phase space available for in medium nucleon-nucleon collisions.

It should be mentioned that the onset of superfluidity and/or superconductivity in neutron stars is also possible at the higher densities typical of the core region. In this case, however, pairing predominantly involves nucleons having relative angular momentum $\ell = 1$ and total spin $S = 1$.

A detailed discussion of the structure and properties of the neutron star crust is beyond the scope of this book. The interested reader is referred to the exhaustive review of Chamel and Hansel et al. [31]. The region of nuclear and supranuclear density, extending over most of the star volume and accounting for over 90% of its mass, will be discussed in the following chapter.

---

[2] According to the standard spectroscopic notation, the two-nucleon state $^1S_0$ corresponds to orbital angular momentum $\ell = 0$, spin $S = 0$ and total angular momentum $J = 0$. The total isospin of this state, dictated by the requirement of antisymmetry under particle exchange, is $T = 1$.

The Neutron Star Core

3

Abstract

At densities in the range $\varrho_0 \lesssim \varrho \lesssim 2\varrho_0$, with ϱ_0 being the nuclear density, neutron star matter is a uniform fluid of neutrons, with a small admixture of protons and leptons. This chapter, after reviewing the constraints on the EOS of cold nuclear matter that can be inferred from nuclear data, provides a concise summary of the empirical information on the underlying microscopic dynamics. The models of the nuclear Hamiltonian—providing the basis for the description of neutron star matter within non relativistic nuclear many-body theory—as well as the most employed computational approaches are reviewed. The alternative relativistic approach, based on the formalism of quantum field theory and mean field approximation, is also described.

3.1   Preamble

Understanding the properties of matter at densities comparable to the central density of atomic nuclei, ϱ_0, is made difficult by *both* the complexity of the interactions *and* the approximations necessarily implied in the theoretical treatment of quantum mechanical many-particle systems. The picture becomes even more problematic in the region of *supranuclear* density, corresponding to $\varrho > \varrho_0$, as the available empirical information is scarce, and theoretical models largely rely on a mixture of extrapolation and speculation.

The approach based on non relativistic quantum mechanics and phenomenological nuclear Hamiltonians, while allowing a good description of nuclear bound states and nucleon-nucleon scattering data, fails to fulfill the constraint of causality, because it leads to predict a speed of sound in matter that exceeds the speed of light at large density. On the other hand, the approach based on relativistic quantum field theory, while fulfilling the requirement of causality by construction, assumes

© The Author(s), under exclusive license to Springer Nature Switzerland AG 2023
O. Benhar, *Structure and Dynamics of Compact Stars*, Lecture Notes
in Physics 1019, https://doi.org/10.1007/978-3-031-35628-5_3

a somewhat oversimplified dynamics, not constrained by the observed properties
of the two- and three-nucleon system. In addition, it is plagued by the uncertainty
inherent in the use of the mean field approximation, which is long known to fail in
strongly correlated systems, such as, for example, the van der Waals fluid discussed
in Sect. 1.4 [32].

3.2   Constraints on the Nuclear Matter EOS

The body of data on nuclear masses can be used to constrain the density dependence
predicted by theoretical models of uniform nuclear matter at zero temperature. Note
that the zero temperature limit is fully justified, as the typical temperature of the
neutron star interior is $\sim 10^9$ K ~ 0.1 MeV, to be compared to typical nucleon Fermi
energies of tens of MeV.

The A-dependence of the nuclear binding energy, discussed in Sect. 2.2.1, is
well described by the semi empirical mass formula (2.26) . In the limit of large A,
and neglecting the effect of Coulomb repulsion between protons, the only surviving
contribution for $Z = A/2$ is the term linear in A. Hence, the coefficient $a_V \approx$
-16 MeV can be interpreted as the binding energy per nucleon of *isospin-symmetric
nuclear matter* (SNM), an ideal uniform system consisting of equal number of
protons and neutrons coupled by strong interactions only. The equilibrium density
of such a system can be inferred exploiting saturation of nuclear densities, that is,
the observation that the charge density at the center of atomic nuclei—accurately
measured by elastic electron-nucleus scattering experiments [33]—becomes nearly
independent of A for large A. Figure 3.1, showing the radial dependence of the
charge number density normalised to A for $A = 16, 58, 116$, and 208, indicates that
the nucleon density of SNM at equilibrium is $n_0 \approx 0.16$ fm^{-3}.

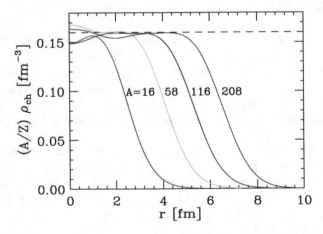

Fig. 3.1 Saturation of charge densities at the center of nuclei with $A = 16, 58, 116$, and 208,
measured by elastic electron-nucleus scattering experiments

In conclusion, the empirical equilibrium properties of SNM inferred from nuclear data turn out to be

$$\left(\frac{E}{A}\right)_{n=n_0} = -16 \text{ MeV}, \quad n_0 \approx 0.16 \text{ fm}^{-3} . \tag{3.1}$$

In the vicinity of the equilibrium density, the energy per nucleon $e = E/A$ can be expanded according to

$$e(n) \approx e_0 + \frac{1}{2} \frac{K}{9} \frac{(n - n_0)^2}{n_0^2} , \tag{3.2}$$

where

$$K = 9 n_0^2 \left(\frac{\partial^2 e}{\partial n^2}\right)_{n=n_0} = 9 \left(\frac{\partial P}{\partial n}\right)_{n=n_0} \tag{3.3}$$

is the (in)compressibility module, that can be extracted from the measured excitation energies of nuclear vibrational states. Owing to the difficulties implied in the interpretation of these experiments, however, empirical estimates of K have a rather large uncertainty. Recent global analyses of the available data yield $K = 240 \pm 20$ MeV [34, 35]. Note that *lower* (*higher*) values of K correspond to *softer* (*stiffer*) EOS.

The quadratic extrapolation of Eq. (3.2) cannot be expected to work far from equilibrium density. To see this, consider that the assumption of parabolic behaviour of $e(n)$ at $n \gg n_0$ leads to predict a speed of sound in matter—defined by Eq. (1.45)—larger than the speed of light, that is, to

$$c_s^2 = \frac{1}{n} \left(\frac{\partial P}{\partial e}\right) > 1 , \tag{3.4}$$

independent of the value of K.

Equation (3.4) implies that causality requires

$$\left(\frac{\partial P}{\partial \epsilon}\right) < 1 , \tag{3.5}$$

ϵ being the matter energy density. For a relativistic Fermi gas $\epsilon \propto n^{4/3}$, and

$$P \leq \frac{\epsilon}{3} , \quad c_s^2 \leq \frac{1}{3} . \tag{3.6}$$

with the upper bound corresponding to massless fermions. In the presence of inter-actions, however, the above limit can be easily exceeded. For example, describing repulsive nucleon-nucleon interactions in terms of a rigid core obviously leads to predict infinite pressure at finite density.

The relation between microscopic dynamics and the speed of sound in matter has been analysed by Zel'dovich in the early 1960s [36]. The results of this study, performed using the formalism of relativistic quantum field theory, shows that the causality limit, corresponding to $c_s^2 = 1$, is in fact be reached in a simple semi realistic model of nuclear dynamics, in which nucleons are described as scalar particles interacting through exchange of a vector meson. A detailed description of Zel'dovich's model can be found in Appendix 1.

3.3   Microscopic Models of the Nuclear Matter EOS

In addition to imposing important constraints on the nuclear matter EOS, the systematics of measured nuclear properties provides important information on the forces acting among nucleons, thus shedding light on the underlying microscopic dynamics. This chapter is devoted to a description of the dynamical models employed in theoretical studies of neutron stars, as well as an overview of the employed computational approaches.

3.3.1   Empirical Information on Nuclear Forces

The saturation of the charge density distributions and the observation that the binding energy per nucleon becomes largely independent of the mass number, A, for $A \geq 20$—illustrated in Figs. 3.1 and 2.5, respectively—reveal two prominent feature of nuclear dynamics: nucleon-nucleon forces are strongly repulsive at short distances and have finite range $r_0 \ll R_A$, R_A being the nuclear radius.

In addition, the observation that the spectra of mirror nuclei—that is, pairs of nuclei with the same A and charges differing by one unit—exhibit remarkable similarities indicates that proton and neutrons have nearly identical interactions. This property, referred to as *charge symmetry* of nuclear forces, reflects a more general symmetry: the *isotopic invariance*.

Neglecting the $\sim$0.1% mass difference, neutron and proton can be treated as a single particle, the *nucleon*, that can be found in two different states specified by a quantum number dubbed *isospin*. It follows that protons and neutrons can be described using the same formalism developed for the electron spin. The nucleon is an isospin doublet, and proton and neutron correspond to isospin projections $+1/2$ and $-1/2$, respectively. A two-nucleon state is specified the total isospin, T, and its projection, T_3. Note that two protons and two neutrons always have $T = 1$, while a proton-neutron pair may have either $T = 0$ of $T = 1$. Isospin invariance dictates that the force acting between two nucleons depends on T only, not on T_3.

3.3.2   The Nucleon-Nucleon Interaction

The details of the nucleon-nucleon (NN) interaction can be best studied in the two-nucleon system. There is *only one* NN bound state, the nucleus of deuterium, or deuteron, ^{2}H, consisting of a proton and a neutron coupled to total spin and isospin $S = 1$ and $T = 0$. This is a clear manifestation of the *spin-isospin dependence* of nuclear forces.

Another important piece of information can be deduced from the observation that the deuteron has non vanishing electric quadrupole moment, implying that its charge distribution is not spherically symmetric. It follows that the NN interaction is *not spherically symmetric*.

Besides the properties of the two-nucleon bound state, the large data base of phase shifts measured in NN scattering experiments—comprising ~5000 data points corresponding to energies up to 350 MeV in the lab frame—provides valuable additional information on the nature of NN forces.

The theoretical description of the NN interaction was first attempted by Yukawa in 1935 [37]. He made the hypothesis that nucleons interact through the exchange of a particle, whose mass μ can be deduced from the interaction range r_0 using

$$r_0 \sim \frac{1}{\mu} . \tag{3.7}$$

For $r_0 \sim 1$ fm, the above relation yields $\mu \sim 200$ MeV.

Yukawa's suggestion was successfully implemented identifying the exchanged particle with the π-meson, or *pion*, discovered in 1947, the mass of which turned out to be $m_\pi \sim 140$ MeV. Experiments have shown that the pion is a spin-zero particle with negative intrinsic parity. The spin has been deduced from the balance of the reaction

$$\pi^+ + ^2\text{H} \leftrightarrow p + p , \tag{3.8}$$

while the intrinsic parity has been determined by observing π^- capture from the K-shell of the deuterium atom

$$d + \pi^- \rightarrow n + n , \tag{3.9}$$

leading to the appearance of two neutrons. The π-meson is observed in three charge states, denoted π^+, π^- and π^0, and can be described using the isospin formalism as a T=1 triplet, the charge states being associated with isospin projections $T_3 = +1$, 0, and -1, respectively.

The simplest π-nucleon coupling compatible with the observation that nuclear interactions conserve parity has the pseudoscalar form $ig\gamma^5\tau$ [38], where g is a coupling constant and τ, analog to the spin operator, describes the isospin of the nucleon. With this choice of the interaction vertex, the amplitude of the

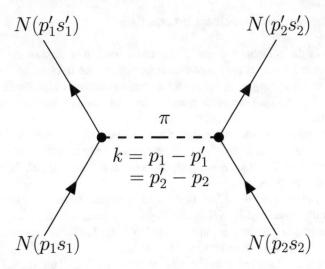

Fig. 3.2 Feynman diagram depicting one-pion-exchange between two nucleons. The corresponding amplitude is given by Eq. (3.10)

process depicted in Fig. 3.2 can be readily written using standard Feynman diagram techniques. The resulting expression reads

$$\langle f|M|i\rangle = -ig^2 \, \frac{\bar{u}(p_2', s_2')\gamma_5 u(p_2, s_2)\bar{u}(p_1', s_1')\gamma_5 u(p_1, s_1)}{k^2 - m_\pi^2} \, \langle \boldsymbol{\tau}_1 \cdot \boldsymbol{\tau}_2 \rangle \,, \qquad (3.10)$$

where $k = p_1' - p_1 = p_2 - p_2'$, $k^2 = k_\mu k^\mu = k_0^2 - |\mathbf{k}|^2$, $u(p, s)$ is the Dirac spinor associated with a nucleon of four momentum $p \equiv (\mathbf{p}, E)$, with $E=\sqrt{\mathbf{p}^2 + m^2}$, and spin projection s. The matrix element in isospin space is denoted

$$\langle \boldsymbol{\tau}_1 \cdot \boldsymbol{\tau}_2 \rangle = \eta_{2'}^\dagger \boldsymbol{\tau} \eta_2 \, \eta_{1'}^\dagger \boldsymbol{\tau} \eta_1 \,, \qquad (3.11)$$

with η_i being the two-component Pauli spinor describing the state of particle i.

In the nonrelativistic limit, Yukawa's theory leads to define a NN interaction potential, that can be written in coordinate space as

$$\begin{aligned}
v_\pi &= \frac{g^2}{4m^2} \, (\boldsymbol{\tau}_1 \cdot \boldsymbol{\tau}_2)(\boldsymbol{\sigma}_1 \cdot \boldsymbol{\nabla})(\boldsymbol{\sigma}_2 \cdot \boldsymbol{\nabla}) \, \frac{e^{-m_\pi r}}{r} \\
&= \frac{g^2}{(4\pi)^2} \frac{m_\pi^3}{4m^2} \frac{1}{3}(\boldsymbol{\tau}_1 \cdot \boldsymbol{\tau}_2) \left\{ \left[(\boldsymbol{\sigma}_1 \cdot \boldsymbol{\sigma}_2) + S_{12}\left(1 + \frac{3}{x} + \frac{3}{x^2}\right) \right] \frac{e^{-x}}{x} \right. \\
&\qquad\qquad \left. - \frac{4\pi}{m_\pi^3}(\boldsymbol{\sigma}_1 \cdot \boldsymbol{\sigma}_2)\delta^{(3)}(\mathbf{r}) \right\} \,, \qquad (3.12)
\end{aligned}$$

where $x = m_\pi |\mathbf{r}|$ and

$$S_{12} = \frac{3}{r^2}(\boldsymbol{\sigma}_1 \cdot \mathbf{r})(\boldsymbol{\sigma}_2 \cdot \mathbf{r}) - (\boldsymbol{\sigma}_1 \cdot \boldsymbol{\sigma}_2) , \tag{3.13}$$

is reminiscent of the operator describing the noncentral interaction between two magnetic dipoles.[1] A detailed derivation of Eq. (3.12) can be found in Appendix 2.

For $g^2/4\pi \approx 14$, the one-pion-exchange (OPE) potential provides a description of the long range part of the NN interaction, corresponding to $|\mathbf{r}| > 1.5$ fm. This information is provided by the fit of NN scattering phase shifts in states of high angular momentum, in which—due to the presence of a strong centrifugal barrier—the probability of finding the two interacting nucleons at small relative distances is negligibly small.

At medium and short range, more complex processes involving the exchange of two or more pions—possibly interacting among themselves—or heavier particles—such as the ρ and ω mesons, the masses of which are 769 and 783 MeV, respectively—should be taken into account. In addition, when their relative distance becomes very small, typically $|\mathbf{r}| \lesssim 0.5$ fm, the nucleons, being composite and finite in size, are expected to overlap. In this regime, NN interactions should in principle be described in terms of interactions between nucleon constituents, that is, quarks and gluons, as predicted by the fundamental theory of strong interactions: *Quantum Chromo-Dynamics*, or QCD [39].

Phenomenological potentials describing the *full* NN interaction are written in the form

$$v = \widetilde{v}_\pi + v_R , \tag{3.14}$$

where $\widetilde{v}_\pi$ is the OPE potential defined by Eqs. (3.12) and (3.13), stripped of the δ-function contribution, whereas v_R describes the interaction at medium and short range. The spin-isospin dependence and the noncentral nature of the NN interactions can be effectively described rewriting Eq. (3.14) in the form

$$v_{ij} = \sum_{ST} \left[v_{TS}(r_{ij}) + \delta_{S1} v_{tT}(r_{ij}) S_{12} \right] P_S \Pi_T , \tag{3.15}$$

where S and T are the total spin and isospin of the interacting pair. The spin-projection operators P_S are defined as

$$P_0 = \frac{1}{4}(1 - \boldsymbol{\sigma}_1 \cdot \boldsymbol{\sigma}_2) , \quad P_1 = \frac{1}{4}(3 + \boldsymbol{\sigma}_1 \cdot \boldsymbol{\sigma}_2) , \tag{3.16}$$

---

[1] Note that the use of pseudo-vector, rather than pseudo-scalar, pion-nucleon coupling—as prescribed by chiral symmetry—leads to the same NN potential in the non relativistic limit.

and satisfy the relations

$$P_0 + P_1 = 1 \, , \quad P_S|S'\rangle = \delta_{SS'}|S'\rangle \, , \quad P_S P_{S'} = P_S \delta_{SS'} \, , \tag{3.17}$$

The isospin projection operators Π_T can be written as in Eq. (3.16) replacing $\boldsymbol{\sigma} \to \boldsymbol{\tau}$. The functions $v_{TS}(r_{ij})$ and $v_{Tt}(r_{ij})$ describe the radial dependence of the interaction in the different spin-isospin channels, and reduce to the corresponding components of the one pion exchange potential at large r_{ij}. Their explicit form involves a set of parameters whose values are determined in such a way as to reproduce the available NN data, that is, deuteron binding energy, charge radius and quadrupole moment, and the NN scattering phase shifts.

Substitution of Eq. (3.16) and the corresponding expressions for the isospin projection operators allows one to rewrite Eq. (3.15) in the form

$$v_{ij} = \sum_{n=1}^{6} v^{(n)}(r_{ij}) O_{ij}^{(n)} \, , \tag{3.18}$$

where

$$O_{ij}^{(n)} = 1, \; (\boldsymbol{\tau}_i \cdot \boldsymbol{\tau}_j), \; (\boldsymbol{\sigma}_i \cdot \boldsymbol{\sigma}_j), \; (\boldsymbol{\sigma}_i \cdot \boldsymbol{\sigma}_j)(\boldsymbol{\tau}_i \cdot \boldsymbol{\tau}_j), \; S_{ij}, \; S_{ij}(\boldsymbol{\tau}_i \cdot \boldsymbol{\tau}_j) \, , \tag{3.19}$$

and the $v^{(n)}(r_{ij})$ are linear combination of the $v_{TS}(r_{ij})$ and $v_{tT}(r_{ij})$. Note that the operators defined in Eq. (3.19) form an algebra. They satisfy the relations

$$O_{ij}^{(n)} O_{ij}^{(m)} = \sum_{\ell} K_{nm\ell} O_{ij}^{(\ell)} \, , \tag{3.20}$$

where the coefficients $K_{nm\ell}$ can be easily obtained from the properties of Pauli matrices. Equations (3.20) can be exploited to greatly simplify the calculation of nuclear observables based on the representation of the NN potential given by Eqs. (3.18) and (3.19).

The typical shape of the NN potential in the state of relative angular momentum $\ell = 0$ and total spin and isospin $S = 0$ and $T = 1$ is shown in Fig. 3.3. The short range repulsive core, to be ascribed to heavy meson exchange or to more complicated mechanisms involving the internal structure of the nucleon, is followed by an intermediate range attractive region, largely due to two-pion exchange processes. Finally, at large interparticle distance the OPE mechanism dominates. Note the similarity with the van der Waals potential of Fig. 1.1.

State-of-the-art phenomenological models of the NN potential, such as the Argonne v_{18} (A18) potential [40], include twelve additional terms to the sum appearing in the right-hand side of Eq. (3.18). The operators corresponding to $n = 7, \ldots, 14$ are associated with non-static contributions, notably spin-orbit terms, while those corresponding to $n = 15, \ldots, 18$ take into account small violations of charge symmetry and charge independence.

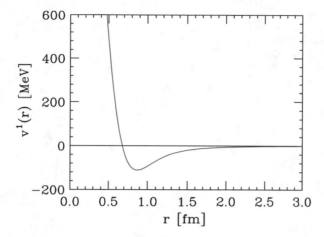

Fig. 3.3 Radial dependence of the NN potential describing the interaction between two nucleons in the state of relative angular momentum $\ell = 0$, and total spin and isospin $S = 0$ and $T = 1$

Over the past two decades, a more elegant and fundamental approach originally proposed by Weinberg [41], in which the NN potential is obtained within the framework of chiral effective field theory [42, 43], has enjoyed a great deal of popularity within the Nuclear Physics community. However, chiral potentials are known to be adequate to provide reliable predictions of nuclear matter properties only up to twice nuclear saturation density [44, 45]. As a consequence, their use in neutron star studies appears to be severely limited.

3.3.3 Irreducible Three-Nucleon Interactions

Phenomenological NN potentials, while explaining the observed properties of the two-nucleon system by construction, fail to accurately predict the ground-state energy of the three-nucleon bound states , the calculation of which can be performed exactly by solving the Faddeev equations [46, 47]. To reconcile theoretical results and data, the nuclear Hamiltonian must be supplemented with a potential describing *irreducible* three-nucleon (NNN) interactions, that is, interaction involving three particles that cannot be reduced to a combination of two-nucleon interactions.

The inclusion of three-body forces is long known to be needed to describe the interactions involving composite systems neglecting their internal structure. For example, a three-body force can be introduced to take into account the effect of the tidal deformation, which explicitly depends on the position of the Moon, on the motion of a satellite orbiting the Earth; see Fig. 3.4.

Fig. 3.4 Schematic
representation of the
irreducible three-body force
in the Earth-Moon-satellite
system, originating from
Earth's tidal deformation

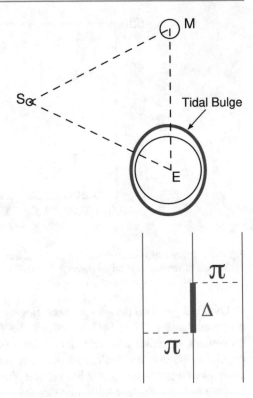

Fig. 3.5 Diagrammatic
representation of a two-pion
exchange process leading to
the appearance of irreducible
NNN forces

Phenomenological models of the NNN potential, such as the Urbana IX (UIX)
potential [48], are generally written in the form

$$V_{ijk} = V_{ijk}^{2\pi} + V_{ijk}^{R} , \qquad (3.21)$$

where $V_{ijk}^{2\pi}$ is an attractive contribution arising from two-pion exchange processes—
originally proposed by Fujita and Miyiazawa in the 1950s [49]—in which one of
the participating nucleons is excited to the spin 3/2 state of mass 1232 MeV dubbed
Δ resonance; see Fig. 3.5. The second term in the right-hand side of Eq. (3.21) is
purely phenomenological, repulsive and independent of isospin. The strengths of the
two contributions to V_{ijk} are determined in such a way as to reproduce the binding
energies of ^{3}He and ^{4}He and the empirical equilibrium density of SNM.

3.3.4 Non Relativistic Nuclear Many-Body Theory

To a remarkably large extent, nuclear systems behave as collections of point-like non relativistic particles, the interactions of which are described by the Hamiltonian

$$H = \sum_{i=1}^{A} \frac{\mathbf{p}_i^2}{2m} + \sum_{j>i=1}^{A} v_{ij} + \sum_{k>j>i=1}^{A} V_{ijk} \,, \tag{3.22}$$

where $\mathbf{p}_i$ is the momentum of the i-th nucleon.

The validity of the assumption that, in spite of their finite size, nucleons can be treated as point-like particles is strongly supported by the analysis of the large body of available electron-nucleus scattering data. Scaling in the variable y— closely related to the momentum of the nucleon participating in the electromagnetic interaction—has been in fact observed by experiments using a broad range of targets, extending from ^{2}H to nuclei as heavy as ^{197}Au [50, 51]. The results of these studies unambiguously demonstrate that in the kinematical region corresponding to momentum transfer $q \gtrsim 1$ GeV and large negative y the beam particles primarily couple to nucleons belonging to correlated pairs, the momentum of which can be in excess of 500 MeV. Based on the analysis of data collected at Jefferson Lab, Subedi et al. [52], argued the appearance of correlated nucleon pairs is associated with large fluctuations of the matter density that can locally be as high as five times the equilibrium density of SNM.

The determination of the nuclear matter EOS requires the solution of the Schrödinger equation

$$H|\Psi_0\rangle = E_0|\Psi_0\rangle \,, \tag{3.23}$$

with H being the Hamiltonian of Eq. (3.22), to obtain the ground-state wave function and energy at any fixed density. Owing to the complexity of the nuclear potentials, however, even the achievement of this limited goal involves severe, in fact insurmountable, difficulties.

The Schrödinger equation can be solved—using either deterministic or stochastic approaches—only for not-too-large A. For $A = 2$ the numerical solution of Eq. (3.23) is straightforward, while for $A = 3$ it can be obtained using the Faddeev formalism. For $4 \leq A \leq 12$, stochastic techniques, such as the Green Function Monte Carlo (GFMC) method, have been successfully employed [53]. The results of these calculations, showing the remarkable accuracy and predictive power of the approach based on non relativistic nuclear many-body theory (NMBT) and the Hamiltonian of Eq. (3.22), are briefly reviewed in the next section.

The Nuclear Many-Body Problem

For $A > 3$ the Scrödinger equation is no longer exactly solvable. The ground-state energy of nuclei having $A \geq 4$ can be estimated exploiting Ritz's variational principle, stating that the expectation value of the Hamiltonian in the *trial* state $|\Psi_V\rangle$ satisfies

$$E_V = \frac{\langle \Psi_V | H | \Psi_V \rangle}{\langle \Psi_V | \Psi_V \rangle} \geq E_0 \,, \tag{3.24}$$

E_0 being the *true* ground state energy. Obviously, the larger the overlap $\langle \Psi_0 | \Psi_V \rangle$ the closer E_V is to E_0.

In principle, E_0 can be estimated performing a functional minimisation of E_V. with a trial ground state chosen in such a way as to embody the effects of interactions. For few-nucleon systems $|\Psi_V\rangle$ it is generally written in the form

$$|\Psi_V\rangle = (1 + U) |\Psi_P\rangle \,, \tag{3.25}$$

with

$$|\Psi_P\rangle = F |\Phi_A(JJ_3TT_3)\rangle \,. \tag{3.26}$$

In the above equations $|\Phi_A(JJ_3TT_3)\rangle$ is a state describing A independent particles coupled to total angular momentum J and total isospin T, with projections J_3 and T_3, while the operators F and U take into account the correlation structure induced by the two- and three-nucleon potentials, respectively. The most prominent correlation effects, associated with the NN potential v_{ij}, are described by the operator F, defined as

$$F = \mathcal{S} \prod_{j>i=1}^{A} f_{ij} \,, \tag{3.27}$$

with the NN correlation, f_{ij}, reflecting the operator structure of the potential of Eqs. (3.18) and (3.19), given by

$$f_{ij} = \sum_{n=1}^{6} f^{(n)}(r_{ij}) O_{ij}^{(n)} \,. \tag{3.28}$$

Note that, because, in general, $\left[f_{ij}, f_{ik} \right] \neq 0$, the product of correlations must be symmetrised through the action of the operator $\mathcal{S}$.

The shape of the radial correlation functions $f^{(n)}(r_{ij})$ is determined by minimising the ground-state energy expectation value E_V. In few nucleon systems this procedure is implemented choosing suitable analytical expressions involving a set adjustable parameters. The main features of the $f^{(n)}(r_{ij})$ are dictated by the behaviour of the corresponding component of the potential v_{ij}. For example, the presence of the largely state-independent repulsive core requires that $f^{(n)}(r_{ij}) \ll 1$, independent of n, at $r \lesssim 1$ fm. On the other, the longer range of $f^{(6)}(r_{ij})$ reflects the

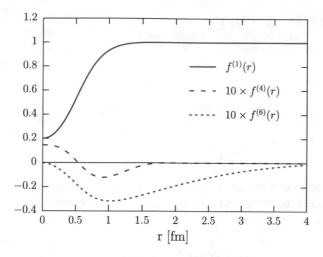

Fig. 3.6 Radial dependence of the NN correlation functions $f^{(n)}$ of Eq. (3.28) corresponding to $n = 1$, 4, and 6. The numerical results have been obtained from minimisation of the ground state energy of SNM at density $n_B = 0.16$ fm^{-3}, with a phenomenological nuclear Hamiltonian including NN and NNN potentials

OPE tail. A representative set of radial correlation functions of SNM at equilibrium density, is shown in Fig. 3.6.

The operator describing irreducible three-nucleon correlations is written in the form

$$U = \sum_{\text{cyclic}} F_{ijk} \, , \tag{3.29}$$

where the sum is extended to all cyclic permutations of the indices, and the F_{ijk} are defined similarly to the f_{ij} of Eq. (3.28).

The difficulty of the variational approach lies in the calculation of the expectation value E_V, which involves a non factorisable integration over $3A$ space coordinate as well as a sum over the spin-isospin degrees of freedom. The issue of dimensionality can be easily grasped writing the variational state in the form

$$|\Psi_V\rangle = \sum_{m=1}^{M} \Psi_m(R)|m\rangle \, , \tag{3.30}$$

where the sum includes all possible spin-isospin states, labelled by the index m, and $R \equiv \{\mathbf{r}_1, \ldots, \mathbf{r}_A\}$ specifies the configuration of the system in coordinate space. For example, in the case of ^{3}He, having $J = T = 1/2$, one finds

$$|1\rangle = |\uparrow p \ \uparrow n \ \downarrow n \rangle$$
$$|2\rangle = |\downarrow p \ \uparrow n \ \downarrow n \rangle$$
$$|3\rangle = |\downarrow n \ \uparrow p \ \downarrow n \rangle$$
$$\ldots = \ldots\ldots ,\tag{3.31}$$

The possible spin states of A spin-1/2 particles are 2^A and, since Z of the A nucleons can be protons, there are $A!/Z!(A - Z)!$ isospin states. Hence, the sum over m in Eq. (3.30) involves

$$M = 2^A \ \frac{A!}{Z!(A - Z)!} ,\tag{3.32}$$

contributions.

In the representation of Eq. (3.30) the nuclear Hamiltonian H is a $M \times M$ matrix whose elements depend upon R. To obtain E_V one has to evaluate the $M \times M$ integrals

$$\int dR \Psi_n^\dagger(R) H_{nm} \Psi_m(R) ,\tag{3.33}$$

using Monte Carlo techniques.

The expectation value of any operator O in the state Ψ_V can be written in the form

$$\langle O \rangle = \sum_{m,n} \int dR \left[\frac{\Psi_m^\dagger(R) O_{mn}(R) \Psi_n(R)}{P_{mn}(R)} \right] P_{mn}(R)$$
$$= \sum_{m,n} \int dR \ \tilde{O}_{mn}(R) P_{mn}(R) ,\tag{3.34}$$

where

$$\tilde{O}_{mn} = \frac{\Psi_m^\dagger(R) O_{mn}(R) \Psi_n(R)}{P_{mn}(R)} ,\tag{3.35}$$

and the *probability distribution* $P_{mn}(\mathbf{R})$ is given by

$$P_{mn}(R) = |\text{Re} \ (\Psi_m^\dagger(R) \Psi_n(R))| .\tag{3.36}$$

Let $\{R_p\} \equiv \{R_1, \ldots, R_{N_c}\}$ be a set of N_c configurations drawn from the probability distribution of Eq. (3.34), that is, such that the probability that a configuration $\tilde{R}$ belongs to the set $\{R_p\}$ is proportional to $P_{mn}(R)$. It then follows that

$$\int dR \, \tilde{O}_{mn}(R) P_{mn}(R) = \lim_{N_c \to \infty} \frac{1}{N_c} \sum_{p=1}^{N_c} \tilde{O}_{mn}(R_p) . \tag{3.37}$$

The above scheme, referred to as Variational Monte Carlo (VMC) method, allows one to obtain estimates of the ground state energy E_0, the accuracy of which is limited by the statistical error associated with the use of a finite configuration set and by the uncertainty associated with a specific choice of the trial wave function. The second source of error can be removed using the Green Function Monte Carlo (GFMC) approach.

Let $\{|\Psi_m\rangle\}$ be the complete set of eigenstates of the nuclear Hamiltonian, satisfying

$$H|\Psi_m\rangle = E_m|\Psi_m\rangle . \tag{3.38}$$

The trial variational wave function can be expanded according to

$$|\Psi_V\rangle = \sum_n \beta_n |\Psi_m\rangle , \tag{3.39}$$

implying

$$\lim_{\tau \to \infty} e^{-H\tau}|\Psi_V\rangle = \lim_{\tau \to \infty} \sum_n \beta_n \, e^{-E_m\tau}|\Psi_m\rangle = \beta_0 \, e^{-E_0\tau}|\Psi_0\rangle . \tag{3.40}$$

The above equation shows that the evolution of the variational ground state to infinite imaginary time projects out the *true* ground state of the nuclear Hamiltonian, and allows to pin down the corresponding eigenvalue.

Numerical calculations are carried out by dividing the imaginary time interval τ in N steps of length $\Delta\tau = \tau/N$ and rewriting

$$e^{-H\tau} = \left(e^{-H\Delta\tau}\right)^N . \tag{3.41}$$

The state at imaginary time $(i+1)\Delta\tau$ is can be obtained from the one corresponding to $\tau = i\Delta\tau$ using the relation

$$|\Psi_V^{i+1}\rangle = e^{-H\Delta\tau}|\Psi_V^i\rangle , \tag{3.42}$$

that can be rewritten

$$\langle R'S'T'|\Psi_V^{i+1}\rangle = \sum_{ST} \int dR \, \langle R'S'T'|e^{-H\Delta\tau}|RST\rangle\langle RST|\Psi_V^i\rangle \,, \tag{3.43}$$

or

$$\Psi_{V,S'T'}^{i+1}(R) = \sum_{ST} \int dR \, G_{S'T',ST}(R', R)\Psi_{V,ST}^i(R) \,, \tag{3.44}$$

where $|RST\rangle$ specifies the configuration of the system in coordinate, spin and isospin space. The Green's function appearing in the above equation, yielding the amplitude for the system to evolve from $|RST\rangle$ to $|R'S'T'\rangle$ during the imaginary time interval $\Delta\tau$, is defined as

$$G_{S'T',ST}(R', R) = \langle R'S'T'|e^{-H\Delta\tau}|RST\rangle \,. \tag{3.45}$$

The GFMC approach has been successfully employed to describe the ground and low-lying excited states of nuclei having A up to 12. As an example, the results of calculations for $A \leq 8$, carried out using a nuclear Hamiltonian comprising the Argonne v_{18} NN potential and the UIX NNN potential, are summarised in Table 3.1 and Fig. 3.7. It clearly appears that the non relativistic approach—based on a dynamics modelled to reproduce the properties of two- and three-nucleon systems—has a remarkable predictive power.

3.3.5   Nuclear Matter Theory

In the case of neutron stars, having $A \sim 10^{57}$, the application of the computational techniques described in the previous section is not straightforward. VMC calculations can be performed modelling nuclear matter as a collection of nucleons

Table 3.1 Experimental and quantum Monte Carlo ground-state energies of nuclei with $2 \leq A \leq 8$. The columns marked VMC, GFMC and Expt. show the variational Monte Carlo, Green Function Monte Carlo, and experimental results, respectively. Adapted from Ref. [54] with permissions, © APS 2000. All rights reserved

$^AZ(J^\pi; T)$	VMC	GFMC	Expt
^{2}H(1^+; 0)	-2.2248		-2.2246
^{3}H($(1/2)^+$; 1/2)	-8.15	-8.47	-8.48
^{4}He(0^+; 0)	-26.97	-28.34	-28.30
^{6}He(0^+; 1)	-23.64	-28.11	-29.27
^{6}Li(1^+; 0)	-27.10	-31.15	-31.99
^{7}He($(3/2)^-$; 3/2)	-18.05	-25.79	-28.82
^{7}Li($(3/2)^-$; 1/2)	-31.92	-37.78	-39.24
^{8}He(0^+; 2)	-17.98	-27.16	-31.41
^{8}Li(2^+; 1)	-28.00	-38.01	-41.28
^{8}Be(0^+; 0)	-45.47	-54.44	-56.50

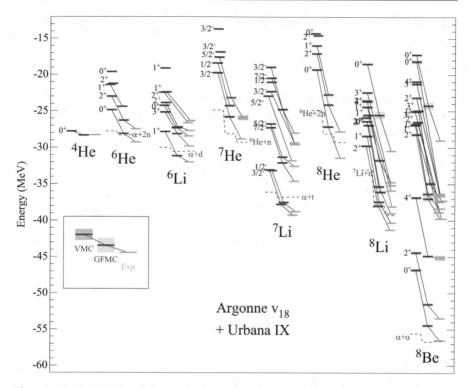

Fig. 3.7 VMC, GFMC, and experimental energies of the ground and low-lying excited states of nuclei with $2 \le A \le 8$. The light shaded area shows either the Monte Carlo statistical error or the experimental uncertainty. Reprinted from Ref. [54] with permissions, © APS 2000. All rights reserved

confined to a periodic box, but numerical studies have so far been limited to a system of 14 neutrons. A more advanced technique referred to as Auxiliary Field Diffusion Monte Carlo (AFDMC), in which the configurations of the system are sampled in both coordinate and spin-isospin space, has allowed to handle up to 132 neutrons [55]. The approach based on the Monte Carlo method, while being very powerful and potentially very accurate, is lacking the flexibility needed to carry out calculations of neutron matter properties other than the energy per particle, which play an important role in determining neutron star properties. In many instances, more approximate methods appear to be better suited for astrophysical applications.

The simplest approximation scheme is based on the replacement of the NN and NNN potentials appearing in Eq. (3.22) with a *mean field*, which amounts to substituting

$$\sum_{j>i=1}^{A} v_{ij} + \sum_{k>j>i=1}^{A} V_{ijk} \rightarrow \sum_{i=1}^{A} U_i \,, \qquad (3.46)$$

with the potential U chosen in such a way that the *single particle* Hamiltonian

$$h_i = \frac{p_i^2}{2m} + U_i , \tag{3.47}$$

be diagonalisable. Within this framework the nuclear ground state wave function reduces to a Slater determinant, constructed using the A lowest energy eigenstates of the operator h_i

$$|\Psi_0\rangle = \frac{1}{\sqrt{A!}}\det\{\phi_i\} , \tag{3.48}$$

with the ϕ_i's ($i = 1, 2,\ldots, A$) being solutions of the Schrödinger equation

$$h_i|\phi_i\rangle = \epsilon_i|\phi_i\rangle . \tag{3.49}$$

The corresponding ground state energy is given by

$$E_0 = \sum_{i=1}^{A} \epsilon_i . \tag{3.50}$$

This procedure is the basis of the nuclear shell model, that has been successfully applied to explain many nuclear properties; see, e.g., Ref. [56].

Matter in the neutron star interior, however, is a uniform, dense nuclear fluid, whose single particle wave functions are known to be plane waves, as dictated by translational invariance. Shell effects are not expected to play a major role in such a system. On the other hand, the strong correlations between nucleons induced by the NN potential, not taken into account within the mean field approximation, become more and more important as the density increases, and must be properly taken into account.

Let us first consider the case of SNM. Neglecting, for the sake of simplicity, three-nucleon forces, the nuclear matter Hamiltonian can be written as in Eq. (3.22) with v_{ij} denoting the NN potential. In the absence of interactions, the wave function is a Slater determinant of single particle states

$$\varphi_{\mathbf{k}\sigma\tau}(\mathbf{r}) = \frac{1}{\sqrt{V}}e^{\mathbf{k}\cdot\mathbf{r}} \chi_\sigma \eta_\tau , \tag{3.51}$$

with χ and η being Pauli spinors in spin and isospin space, respectively, and $|\mathbf{k}| < k_{F_\tau} = (3\pi^2 n_\tau/2)^{1/3}$. Here n_τ, with $\tau = p, n$, denotes the proton and neutron number density.

The main problem associated with the application of many-body perturbation theory to nuclear matter is the presence of the strongly repulsive core in the NN potential, clearly visible in Fig. 3.3, that makes the matrix elements

$$\langle \varphi_{\mathbf{k}_1' \sigma_1' \tau_1'} \varphi_{\mathbf{k}_2' \sigma_2' \tau_2'} | v_{12} | \varphi_{\mathbf{k}_1 \sigma_1 \tau_1} \varphi_{\mathbf{k}_2 \sigma_2 \tau_2} \rangle \, , \tag{3.52}$$

very large, or even divergent. This difficulty can be circumvented either renormalising the interaction potential or redefining the basis states.

G-Matrix Perturbation Theory

Within the first approach the Hamiltonian is first split in two pieces according to

$$H = H_0 + H_1 \, , \tag{3.53}$$

with

$$H_0 = \sum_{i=1}^{N} (K_i + U_i) \, , \tag{3.54}$$

where $K = -\nabla^2/2m$ is the kinetic energy operator in coordinate space, and

$$H_1 = \sum_{j>i=1}^{N} v_{ij} - \sum_{i=1}^{N} U_i \, , \tag{3.55}$$

with the single particle potential generally chosen in such a way as to make the perturbative expansion rapidly convergent. The interaction Hamiltonian H_1 is then treated perturbatively, summing up an infinite set of diagrams to overcome the problems associated with the calculation of the matrix elements (3.52). This procedure leads to the integral equation defining the G-matrix

$$G(W) = v - v \frac{Q}{W} G \, . \tag{3.56}$$

The G-matrix, schematically represented by the diagrams of Fig. 3.8, is the operator describing NN scattering in the nuclear medium. The quantity W appearing in Eq. (3.56) is the energy denominator associated with the propagator of the intermediate state, while the operator Q prevents scattering to states within the Fermi sea, forbidden by Pauli exclusion principle.

The state describing two interacting nucleons, ψ_{ij}, can be written in terms of G using the Bethe-Goldstone equation

$$\psi_{ij} = \phi_{ij} - \frac{Q}{W} G \, \psi_{ij} \, , \tag{3.57}$$

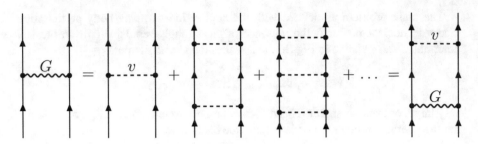

Fig. 3.8 Diagrammatic representation of Eq. (3.56)

with $\phi_{ij} = \mathcal{A}\varphi_i\varphi_j$, where $\varphi_i = \varphi_{\mathbf{k}_i\sigma_i\tau_i}$ is the corresponding unperturbed state, see Eq. (3.51), and $\mathcal{A}$ is the antisymmetrisation operator. From Eq. (3.57) it follows that the matrix elements of G between unperturbed states

$$\langle \phi_{i'j'}|G|\phi_{ij}\rangle = \langle \phi_{i'j'}|v|\psi_{ij}\rangle \,, \tag{3.58}$$

are well behaved.

Although the expansion in powers of G is still not convergent, the terms in the perturbative series can be arranged in such a way as to obtain a convergent expansion in powers of the quantity

$$\kappa = n \sum_{ij} \int d^3r |\phi_{ij}(r) - \psi_{ij}(r)|^2 \,, \tag{3.59}$$

where the sum is extended to all states belonging to the Fermi sea. The above definition shows that κ measures the average distortion of the two-nucleon wave function produced by NN forces.

Nuclear matter studies carried out within G-matrix perturbation theory are generally performed in the so-called Brueckner-Hartree-Fock (BHF) approximation, which amounts to only including contributions of order κ to the energy per nucleon. The result of calculations in which terms of order κ^2 are also taken into account indicate that the convergence of the expansion strongly depends on the choice of the single particle potential U_i [57].

CBF Perturbation Theory

Within the alternative scheme, referred to as Correlated Basis Functions (CBF) perturbation theory, the non perturbative effects originating from the short-range repulsive core of the NN potential are incorporated in a set of basis functions [58]. The unperturbed Fermi gas states $|n_{FG}\rangle$ are replaced by *correlated* states, defined as

$$|n\rangle = \frac{F|n_{FG}\rangle}{\langle n_{FG}|F^\dagger F|n_{FG}\rangle^{1/2}} \,, \tag{3.60}$$

where F is the *correlation operator* defined as in Eq. (3.27). The correlated states (3.60) form a complete set but are *not* orthogonal to one another. However, they can be orthogonalised using standard techniques of many-body perturbation theory.

The radial shapes of the $f^{(n)}(r)$, illustrated in Fig. 3.6, are determined minimising the expectation value $E_V = \langle 0|H|0 \rangle$. In nuclear matter, this procedure is carried out through functional minimisation, leading to a set of Euler-Lagrange equations to be solved with the boundary conditions

$$
\lim_{r \to \infty} f^{(n)}(r) = \begin{cases} 1 & n = 1 \\ 0 & n > 1 \end{cases} .
\tag{3.61}
$$

The short range behaviour of the two-nucleon correlation functions is such that the quantity

$$
f_{ij}^{\dagger} H_{ij} f_{ij} = f_{ij}^{\dagger} \left(\frac{\mathbf{p}_i^2}{2m} + \frac{\mathbf{p}_j^2}{2m} + v_{ij} \right) f_{ij} ,
\tag{3.62}
$$

which reduces to H_{ij} at large interparticle distances, is well behaved as $r \to 0$.

Once the correlated basis has been defined, the nuclear Hamiltonian can be split in two pieces as in Eq. (3.53), with H_0 and H_1 being now defined as the diagonal and off-diagonal part of H in the correlated basis, respectively. It follows that

$$
\langle m|H_0|n \rangle = \delta_{mn} \langle m|H|n \rangle
\tag{3.63}
$$

$$
\langle m|H_1|n \rangle = (1 - \delta_{mn}) \langle m|H|n \rangle .
\tag{3.64}
$$

Note that the zero-th order of the CBF expansion of the ground-state energy coincides with the variational estimate E_V. If the two-body correlation function is cleverly chosen, that is, if $E_V \approx E_0$, correlated states have large overlaps with the true eigenstates of the nuclear Hamiltonian, and the matrix elements of H_1 are small. Hence, the perturbative expansion in powers of H_1 turns out to be rapidly convergent.

As pointed out above, the treatment of matrix elements of H between correlated states involves prohibitive difficulties, arising from both the non factorisable multidimensional integrations and the sums over the spin-isospin degrees of freedom. Explicit calculations are performed expanding the matrix element in a series whose terms represent the contributions of subsystems (clusters) containing an increasing number, $N = 2, 3, \ldots, A$, of nucleons. The diagrams representing the terms of the series can be classified according to their topological structure and summed up to all orders solving a set of coupled integral equations, referred to as Fermi Hyper-Netted Chain Single Operator Chain (FHNC/SOC) equations [59, 60].

The Equation of State of Akmal Pandharipande and Ravenhall

The FHNC/SOC summation scheme has been extensively employed to carry out variational calculations of nuclear matter [61–63]. Figures 3.9 and 3.10 show the results reported in the classic paper of Akmal et al. [63]. These studies have been carried out using the A18+UIX Hamiltonian, with and without the relativistic boost correction δv, to be discussed below, the inclusion of which leads to a sizeable softening of the repulsive NNN force. The corresponding potential is labelled UIX'. For comparison, the results of FHNC/SOC calculations performed using a somewhat simplified dynamical model, based on the Urbana v_{14} NN potential, supplemented with a density-dependent potential meant to account for many-nucleon interactions [64], are also shown. The predictions of this model are labelled U14+DDI (FP).

The A18 + δv + UIX' results for SNM also include a density-dependent estimate of the CBF perturbative corrections to the variational ground-state energy. This correction, adjusted to reproduce the empirical saturation properties, exhibits a maximum of 4.5 MeV—corresponding to ∼30% of the interaction energy—located at subnuclear density, $\varrho = 0.11$ fm^{-3}, and rapidly decreases to become negligible in the high-density region relevant to neutron stars. In the literature, the Hamiltonian including the boost-corrected A18 NN potential and the UIX' NNN potential is often referred to as the APR model.

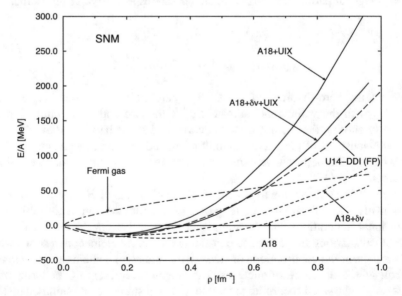

Fig. 3.9 Energy per particle of SNM as a function of baryon number density. The results have been obtained using the variational FHNC/SOC formalism and different nuclear Hamiltonians, with and without inclusion of relativistic boost correction to the NN potential; the meaning of the labels is explained in the text. The prediction of the Fermi gas model is also shown, for comparison. Reprinted from Ref. [63] with permissions, © APS 1998. All rights reserved

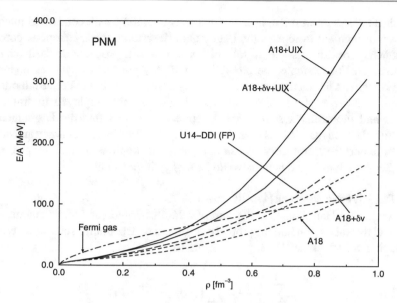

Fig. 3.10 Same as in Fig. 3.9, but for PNM. Reprinted from Ref. [63] with permissions, © APS 1998. All rights reserved

A remarkable feature of the results of Figs. 3.9 and 3.10 is the size of relativistic boost corrections, which are clearly visible at $\varrho \gtrsim 2\varrho_0$ and become large at higher densities.

3.3.6 Relativistic Approaches

The theoretical framework described in the previous section is based on the assumption that the degrees of freedom associated with the carriers of the NN interaction can be eliminated in favour of an instantaneous NN potential. While this procedure has proved remarkably successful at $\rho \sim \rho_0$, as the matter density—and therefore the nucleon Fermi momentum—increases, the relativistic propagation of the nucleons, as well as the retarded propagation of the virtual meson fields giving rise to nuclear forces, are expected to become more and more significant.

In principle, relativistic quantum field theory provides a well defined theoretical framework in which relativistic effects can be taken into account in a fully consistent fashion. Due to the complexity and non perturbative nature of the strong interaction, however, the ab initio approach to the nuclear many problem, based on the QCD Lagrangian, involves prohibitive difficulties. In fact, even the structure of individual hadrons, such as the proton or the π-meson, is not yet understood at a fully quantitative level in terms of the elementary QCD degrees of freedom, let alone the structure of highly condensed hadronic matter at supranuclear densities.

It has to be kept in mind, however, that when dealing with condensed matter it is often convenient to replace the Lagrangian describing the interactions between elementary constituents—be it solvable or not—with properly devised *effective interactions*. For example, the properties of highly condensed systems bound by electromagnetic interactions can be accurately explained using effective interatomic potentials, such as the Lennard-Jones potential, widely employed in studies of atomic and molecular systems [65]. In spite of the fact that the Lagrangian of quantum electrodynamics has been determined to extraordinary accuracy and can be treated in perturbation theory, its use in explicit calculations of the bulk properties of systems such as liquid helium would be highly impractical.

The Relativistic Nuclear Hamiltonian
The relativistic Hamiltonian is the sum of relativistic nucleon kinetic energies and two- and three-body potentials, properly corrected to account for relativistic effects. Replacing *only* the kinetic energy according to

$$\sum_{i=1}^{A} \frac{\mathbf{p}_i^2}{2m} \rightarrow \sum_{i=1}^{A} \left[\sqrt{\mathbf{p}_i^2 + m^2} - m \right] \tag{3.65}$$

does not lead to significant departures from the results of non relativistic calculations of nuclear properties, provided the modified Hamiltonian is employed to obtain the NN potential from a fit the two-nucleon data. Relativistic effects associated with large values of the relative momentum of the interacting nucleons, have been also found to be small, and turn out to be largely canceled by the corrections arising form the replacement of Eq. (3.65).

The most important relativistic correction originates from boosting the NN potential, which is determined in the center-of-mass frame of the NN system. In the presence of the nuclear medium, however, the center-of-mass is *not* at rest, and v_{ij} must be boosted to take into account its motion. The leading boost corrections, arising from the static part of the potential, v_{ij}^s, can be written in the form

$$\delta v_{ij}(\boldsymbol{P}, \boldsymbol{r}) = -\frac{P^2}{8m^2} v_{ij}^s(\boldsymbol{r}) + \frac{(\boldsymbol{P} \cdot \boldsymbol{r})}{8m^2} \boldsymbol{P} \cdot \nabla v_{ij}^s(\boldsymbol{r}) , \tag{3.66}$$

where $\boldsymbol{P} = \mathbf{p}_i + \mathbf{p}_j$ denotes the center-of-mass momentum. A fit to the two-and three-nucleon data and the empirical properties of SNM with the Hamiltonian

$$H_R = \sum_{i} \frac{\mathbf{p}_i^2}{2m} + \sum_{j>i} \left[v_{ij} + \delta v_{ij} \right] + \sum_{k>j>i} V_{ijk}^* , \tag{3.67}$$

shows that the repulsive contribution to the boost corrected NNN potential V_{ijk}^* is reduced by about 30% with respect to the non relativistic counterpart; see, e.g., Ref. [66].

3.3.7 The σ-ω Model

The observation that protons and nucleons largely behave as individual particles interacting through boson exchange suggests that the fundamental degrees of freedom of QCD, quarks and gluons, may in fact, be eliminated in favour of nucleons and mesons, to be regarded as the degrees of freedom of an *effective* field theory. This section is devoted to the description of a simple model in which nuclear matter is viewed as a uniform system of nucleons, described by Dirac spinors, interacting through exchange of a scalar and a vector meson, dubbed, respectively, σ and ω [67].

The basic element of the σ-ω model is the Lagrangian density

$$\mathcal{L} = \mathcal{L}_N + \mathcal{L}_B + \mathcal{L}_{int} , \tag{3.68}$$

where $\mathcal{L}_N$, $\mathcal{L}_B$ and $\mathcal{L}_{int}$ describe free nucleons and mesons and their interactions, respectively. The dynamics of the free nucleon field is dictated by the Dirac Lagrangian

$$\mathcal{L}_N(x) = \bar{\psi}(x)\,(i\slashed{\partial} - m)\,\psi(x) , \tag{3.69}$$

where the nucleon field, denoted by $\psi(x)$, combines the two four-component Dirac spinors describing proton and neutron. The meson Lagrangian is given by

$$\begin{aligned}
\mathcal{L}_B(x) &= \mathcal{L}_\omega(x) + \mathcal{L}_\sigma(x) \\
&= -\frac{1}{4}F^{\mu\nu}(x)F_{\mu\nu}(x) + \frac{1}{2}m_\omega^2 V_\mu(x)V^\mu(x) \\
&\quad + \frac{1}{2}\partial_\mu\phi(x)\partial^\mu\phi(x) - \frac{1}{2}m_\sigma^2\phi(x)^2 ,
\end{aligned} \tag{3.70}$$

where

$$F_{\mu\nu}(x) = \partial_\nu V_\mu(x) - \partial_\mu V_\nu(x) , \tag{3.71}$$

with $V_\mu(x)$ and $\sigma(x)$ being the vector and scalar meson fields, respectively, and m_ω and m_σ the corresponding masses.

In specifying the form of the interaction Lagrangian one requires that, besides being a Lorentz scalar, $\mathcal{L}_{int}(x)$ give rise to a Yukawa-like meson-exchange potential in the static limit. It follows hat

$$\mathcal{L}_{int}(x) = g_\sigma\phi(x)\bar{\psi}(x)\psi(x) - g_\omega V_\mu(x)\bar{\psi}(x)\gamma^\mu\psi(x) , \tag{3.72}$$

where g_σ and g_ω are coupling constants, and the choice of signs is a reminder that the NN interaction comprises both attractive and repulsive contributions.

The equations of motion for the fields follow from the Euler-Lagrange equations associated with the Lagrangian density of Eq. (3.68). The meson fields satisfy

$$(\Box + m_\sigma^2)\phi(x) = g_\sigma \, \bar{\psi}(x)\psi(x), \qquad (3.73)$$

and

$$(\Box + m_\omega^2)V_\mu(x) - \partial_\mu(\partial^\nu V_\nu) = g_\omega \, \bar{\psi}(x)\gamma_\mu\psi(x) \,, \qquad (3.74)$$

while the evolution of the nucleon field is dictated by the equation

$$\left[\left(\not{\partial} - g_\omega\gamma_\mu V^\mu(x)\right) - (m - g_\sigma\phi(x))\right]\psi(x) = 0 \,. \qquad (3.75)$$

The above coupled equations are fully relativistic and Lorentz covariant. However, their solution involves prohibitive difficulties, that can not be circumvented with the use of perturbation theory.

The scheme employed to solve Eqs. (3.73)–(3.75) is known as *relativistic mean field* (RMF) approximation. It essentially amounts to treat $\phi(x)$ and $V_\mu(x)$ as classical fields. The meson field are replaced with their mean values in the ground state of uniform nuclear matter

$$\phi(x) \to \langle\phi(x)\rangle \,, \quad V_\mu(x) \to \langle V_\mu(x)\rangle \,, \qquad (3.76)$$

where $\langle\phi(x)\rangle$ and $\langle V_\mu(x)\rangle$ must be obtained from the equations of motion. In uniform nuclear matter the baryon and scalar densities, $n_B = \langle\psi^\dagger\psi\rangle$ and $n_s = \langle\bar{\psi}\psi\rangle$, as well as the current $j_\mu = \langle\bar{\psi}\gamma_\mu\psi\rangle$, are constants, independent of x. In addition, rotation invariance implies $\langle\bar{\psi}\gamma_i\psi\rangle = 0$, with $i = 1, 2, 3$. As a consequence, the mean values of the meson fields satisfy the relations

$$m_\sigma^2 \langle\phi\rangle = g_\sigma\langle\bar{\psi}\psi\rangle, \qquad (3.77)$$

$$m_\omega^2 \langle V_0\rangle = g_\omega\langle\psi^\dagger\psi\rangle, \qquad (3.78)$$

$$m_\omega^2 \langle V_i\rangle = g_\omega\langle\bar{\psi}\gamma_i\psi\rangle = 0 \,, \quad i = 1, 2, 3 \,. \qquad (3.79)$$

The nucleon equation of motion, rewritten in terms of the mean values of the meson fields, become

$$\left[\left(\not{\partial} - g_\omega\gamma_\mu\langle V^\mu\rangle\right) - (m - g_\sigma\langle\phi\rangle)\right]\psi(x) = 0 \,. \qquad (3.80)$$

In uniform matter, the nucleon states are eigenstates of the four-momentum operator, and can be written as

$$\psi_{\mathbf{k}}e^{ikx} = \psi_{\mathbf{k}}e^{ik_\mu x^\mu} = \psi_{\mathbf{k}}e^{i(k_0 t - \mathbf{k}\cdot\mathbf{r})} \,, \qquad (3.81)$$

with the four-spinors $\psi_\mathbf{k}$ being solutions of

$$\left[\left(\not{k} - g_\omega \gamma_\mu \langle V^\mu \rangle \right) - \left(m - g_\sigma \langle \phi \rangle \right) \right] \psi_\mathbf{k}$$
$$= \left[\gamma_\mu \left(k^\mu - g_\omega \langle V^\mu \rangle \right) - \left(m - g_\sigma \langle \phi \rangle \right) \right] \psi_\mathbf{k} = 0 . \tag{3.82}$$

The above equation can be recast in a form reminiscent of the Dirac equation for a non interacting nucleon. Defining

$$K^\mu = k^\mu - g_\omega \langle V^\mu \rangle , \tag{3.83}$$

$$m^* = m - g_\sigma \langle \phi \rangle , \tag{3.84}$$

one obtains

$$\left(\not{K} - m^* \right) \psi_\mathbf{k} = 0. \tag{3.85}$$

The corresponding energy eigenvalues can be easily obtained using

$$\left(\not{K} + m^* \right) \left(\not{K} - m^* \right) = K_\mu K^\mu - m^{*2} . \tag{3.86}$$

Substitution in Eq. (3.85) yields

$$\left(K_\mu K^\mu - m^{*2} \right) \psi_\mathbf{k} = 0 , \tag{3.87}$$

implying

$$\left(K_\mu K^\mu - m^{*2} \right) = 0 , \tag{3.88}$$

and

$$K_0^2 = (k_0 - g_\omega \langle V_0 \rangle)^2 = |\mathbf{k}|^2 + m^{*2} = E_\mathbf{k}^2 . \tag{3.89}$$

It follows that the energy eigenvalues associated with nucleons and antinucleons can be written

$$e_\mathbf{k} = g_\omega \langle V_0 \rangle + E_\mathbf{k} , \tag{3.90}$$

and

$$\bar{e}_\mathbf{k} = g_\omega \langle V_0 \rangle - E_\mathbf{k} , \tag{3.91}$$

respectively. The above equations give the nucleon and antinucleon energies in terms of the mean values of the meson fields, which are in turn defined in terms of the ground state expectation values of the nucleon densities and current, according to Eqs. (3.77)–(3.79).

The ground state expectation value of any operator of the form $\bar{\psi}\Gamma\psi$ can be evaluated considering that each nucleon state is specified by its momentum, $\mathbf{k}$, and spin-isospin projections. Denoting the average of $\bar{\psi}\Gamma\psi$ in a single particle state by $\langle\bar{\psi}\Gamma\psi\rangle_{\mathbf{k}\alpha}$, where the index α specifies the spin-isospin quantum numbers, the ground state expectation value can be written

$$\langle\bar{\psi}\Gamma\psi\rangle = \sum_\alpha \int \frac{d^3k}{(2\pi)^3}\, \langle\bar{\psi}\Gamma\psi\rangle_{\mathbf{k}\alpha}\, \theta(e_F - e_{\mathbf{k}})\,, \qquad (3.92)$$

where the θ-function restricts the momentum integration to the region corresponding to energies lower than the Fermi energy e_F.

To obtain the single particle average $\langle\bar{\psi}\gamma_\mu\psi\rangle_{\mathbf{k}\alpha}$, we use Eq. (3.85), implying

$$k_0 = \gamma_0\left(\boldsymbol{\gamma}\cdot\mathbf{k} + g_\omega\gamma_\mu\langle V^\mu\rangle + m^*\right)\,. \qquad (3.93)$$

The quantity defined by the above equation can be regarded as a single nucleon Hamiltonian, with eigenvalues given by

$$\langle k_0\rangle_{\mathbf{k}\alpha} = \langle\psi^\dagger k_0\psi\rangle_{\mathbf{k}\alpha} = g_\omega\langle V_0\rangle \pm E_{\mathbf{k}}\,. \qquad (3.94)$$

The ground state expectation value of the baryon density can be readily evaluated from Eqs. (3.93) and (3.94) noting that

$$\frac{\partial}{\partial\langle V_0\rangle}\langle\psi^\dagger k_0\psi\rangle_{\mathbf{k}\alpha} = \frac{\partial}{\partial\langle V_0\rangle}\left(g_\omega\langle V_0\rangle + E_{\mathbf{k}}\right) = g_\omega$$

$$= \langle\psi^\dagger\frac{\partial k_0}{\partial\langle V_0\rangle}\psi\rangle_{\mathbf{k}\alpha} = g_\omega\langle\psi^\dagger\psi\rangle_{\mathbf{k}\alpha}\,, \qquad (3.95)$$

implying

$$\langle\psi^\dagger\psi\rangle_{\mathbf{k}\alpha} = 1\,. \qquad (3.96)$$

It follows that the baryon number density, n_B, can be obtained from Eq. (3.92), leading to

$$n_B = \langle\psi^\dagger\psi\rangle = \nu\int\frac{d^3k}{(2\pi)^3}\theta(e_F - e_{\mathbf{k}})\,, \qquad (3.97)$$

where ν is the degeneracy of the momentum eigenstate, that is, $\nu = 2$ and 4 for PNM and SNM, respectively.

The same procedure can be applied to calculate the ground state expectation value $\langle \bar{\psi} \gamma^i \psi \rangle$, with $i = 1, 2, 3$. Taking the derivative with respect to k_i we find

$$\frac{\partial}{\partial k_i} \langle \psi^\dagger k_0 \psi \rangle_{\mathbf{k}\alpha} = \frac{\partial}{\partial k_i} (g_\omega \langle V_0 \rangle + E_\mathbf{k}) = \frac{\partial E_\mathbf{k}}{\partial k_i}$$

$$= \langle \psi^\dagger \frac{\partial k_0}{\partial k_i} \psi \rangle_{\mathbf{k}\alpha} = \langle \psi^\dagger \gamma^0 \gamma^i \psi \rangle_{\mathbf{k}\alpha} = \langle \bar{\psi} \gamma^i \psi \rangle_{\mathbf{k}\alpha} \ , \quad (3.98)$$

leading to

$$\langle \bar{\psi} \gamma^i \psi \rangle = \nu \int \frac{d^3 k}{(2\pi)^3} \left(\frac{\partial E_\mathbf{k}}{\partial k_i} \right) \theta(e_F - e_\mathbf{k})$$

$$= \frac{\nu}{(2\pi)^3} \int \prod_{j \neq i} dk_j \int dE_\mathbf{k} \, \theta(e_F - e_\mathbf{k}) = 0 \ . \quad (3.99)$$

The above result follows from the observation that, by definition, $e_\mathbf{k} \equiv e_F - g_\omega \langle V_0 \rangle$ everywhere on the boundary of the integration region. The vanishing of the baryon current had been anticipated noting that in uniform matter the mean values of the space components of the vector field vanish, that is, $\langle V_i \rangle = 0$. As a consequence, the energy eigenvalues depend upon the magnitude of the nucleon momentum only, according to

$$e_\mathbf{k} = e_k = g_\omega \langle V_0 \rangle + \sqrt{k^2 + (m - g_\sigma \langle \phi \rangle)^2} \ , \quad (3.100)$$

and the occupied region of momentum space is a sphere. Equation (3.97) then shows that in SNM, with $Z = (A - Z) = A/2$, the baryon density takes the familiar form $n_B = 2k_F^3/(3\pi^2)$, k_F being the Fermi momentum.

Finally, the scalar density $n_s = \langle \bar{\psi} \psi \rangle$ can be evaluated from the derivative of $\langle \psi^\dagger k_0 \psi \rangle_{\mathbf{k}\alpha}$ with respect to m

$$\frac{\partial}{\partial m} \langle \psi^\dagger k_0 \psi \rangle_{\mathbf{k}\alpha} = \frac{\partial E_\mathbf{k}}{\partial m} = \langle \psi^\dagger \frac{\partial k_0}{\partial m} \psi \rangle_{\mathbf{k}\alpha} = \langle \psi^\dagger \gamma_0 \psi \rangle_{\mathbf{k}\alpha} = \langle \bar{\psi} \psi \rangle_{\mathbf{k}\alpha} \ , \quad (3.101)$$

yielding

$$\langle \bar{\psi} \psi \rangle_{\mathbf{k}\alpha} = \frac{(m - g_\sigma \langle \phi \rangle)}{\sqrt{k^2 + (m - g_\sigma \langle \phi \rangle)^2}} \ , \quad (3.102)$$

and

$$\langle \bar{\psi} \psi \rangle = \frac{\nu}{2\pi^2} \int_0^{k_F} k^2 dk \, \frac{(m - g_\sigma \langle \phi \rangle)}{\sqrt{k^2 + (m - g_\sigma \langle \phi \rangle)^2}} \ . \quad (3.103)$$

Collecting together the results of Eqs. (3.97), (3.99) and (3.103), the equations of motion (3.73)–(3.75) can be rewritten in the form

$$g_\sigma \langle \phi \rangle = \left(\frac{g_\sigma}{m_\sigma} \right)^2 \frac{v}{2\pi^2} \int_0^{k_F} \mathbf{k}^2 d|\mathbf{k}| \frac{(m - g_\sigma \langle \phi \rangle)}{\sqrt{\mathbf{k}^2 + (m - g_\sigma \langle \phi \rangle)^2}} , \tag{3.104}$$

$$g_\omega \langle V_0 \rangle = \left(\frac{g_\omega}{m_\omega} \right)^2 n_B = \left(\frac{g_\omega}{m_\omega} \right)^2 v \frac{k_F^3}{6\pi^2} , \tag{3.105}$$

$$m_\omega^2 \langle V_i \rangle = 0 , \quad i = 1, 2, 3 . \tag{3.106}$$

Note that, while Eqs. (3.105) and (3.106) are trivial, Eq. (3.104) implies a self-consistency requirement on the mean value of the scalar field, whose value has to satisfy a transcendental equation.

To obtain the equation of state, that is, the relation between pressure and density—or energy density—of matter, in quantum field theory one starts from the energy-momentum tensor, that for a generic Lagrangian density $\mathcal{L} = \mathcal{L}(\phi, \partial_\mu \phi)$ can be written

$$T^{\mu\nu} = \frac{\partial \mathcal{L}}{\partial(\partial_\mu \phi)} \partial^\nu \phi - g^{\mu\nu} \mathcal{L} , \tag{3.107}$$

$g^{\mu\nu}$ being the metric tensor.

In a uniform system the expectation value of $T^{\mu\nu}$ is directly related to the energy density, ϵ, and pressure, P, through

$$\langle T_{\mu\nu} \rangle = u_\mu u_\nu (\epsilon + P) - g_{\mu\nu} P , \tag{3.108}$$

where u denotes the four-velocity of the system, satisfying $u_\mu u^\mu = 1$. It follows that in the reference frame in which matter is at rest $\langle T_{\mu\nu} \rangle$ is diagonal and

$$\epsilon = \langle T_{00} \rangle = \langle \bar{\psi} \gamma_0 k_0 \psi \rangle - \langle \mathcal{L} \rangle , \tag{3.109}$$

$$P = \frac{1}{3} \langle T_{ii} \rangle = \frac{1}{3} \langle \bar{\psi} \gamma_i k_i \psi \rangle + \langle \mathcal{L} \rangle . \tag{3.110}$$

Within the mean field approximation, the Lagrangian density of the σ-ω model reduces to

$$\mathcal{L}_{MF} = \bar{\psi} \left[i \slashed{\partial} - g_\omega \gamma^0 \langle V_0 \rangle - (m - g_\sigma \langle \phi \rangle) \right] \psi - \frac{1}{2} m_\sigma^2 \langle \phi \rangle^2 + \frac{1}{2} m_\omega^2 \langle V_0 \rangle^2 , \tag{3.111}$$

implying

$$T_{MF}^{\mu\nu} = i\bar{\psi}\gamma^{\mu}\partial^{\nu}\psi - g^{\mu\nu}\left[-\frac{1}{2}m_{\sigma}^2\langle\phi\rangle^2 - \frac{1}{2}m_{\omega}^2\langle V_0\rangle^2\right].\tag{3.112}$$

As a consequence, Eqs. (3.109) and (3.110) become

$$\epsilon = -\langle\mathcal{L}_{MF}\rangle + \langle\bar{\psi}\gamma_0 k_0\psi\rangle,\tag{3.113}$$

$$P = \langle\mathcal{L}_{MF}\rangle + \frac{1}{3}\langle\bar{\psi}\gamma_i k_i\psi\rangle,\tag{3.114}$$

where

$$\langle\bar{\psi}\gamma_0 k_0\psi\rangle = \frac{\nu}{2\pi^2}\int_0^{k_F}|\mathbf{k}|^2 d|\mathbf{k}|\left[\sqrt{\mathbf{k}^2 + (m - g_{\sigma}\langle\phi\rangle)^2} + g_{\omega}\langle V_0\rangle\right]$$

$$= g_{\omega}\langle V_0\rangle n_B + \frac{\nu}{2\pi^2}\int_0^{k_F}|\mathbf{k}|^2 d|\mathbf{k}|\sqrt{\mathbf{k}^2 + (m - g_{\sigma}\langle\phi\rangle)^2}$$

$$= \frac{g_{\omega}^2}{m_{\omega}^2}n_B^2 + \frac{\nu}{2\pi^2}\int_0^{k_F}|\mathbf{k}|^2 d|\mathbf{k}|\sqrt{|\mathbf{k}|^2 + (m - g_{\sigma}\langle\phi\rangle)^2},\tag{3.115}$$

$$\langle\bar{\psi}\gamma_i k_i\psi\rangle = \langle\bar{\psi}(\boldsymbol{\gamma}\cdot\mathbf{k})\psi\rangle = \frac{\nu}{2\pi^2}\int_0^{k_F}d|\mathbf{k}|\frac{|\mathbf{k}|^4}{\sqrt{|\mathbf{k}|^2 + (m - g_{\sigma}\langle\phi\rangle)^2}}.\tag{3.116}$$

Equations (3.115) and (3.116) can be readily obtained using Eqs. (3.94), (3.100), and (3.105), and Eq. (3.99), respectively. Substitution of the above equations into Eqs. (3.113) and (3.114), and use of Eq. (3.111) and the equation of motion for the nucleon field finally yields

$$\epsilon = \frac{1}{2}\frac{m_{\sigma}^2}{g_{\sigma}^2}(m - m^*)^2 + \frac{1}{2}\frac{g_{\omega}^2}{m_{\omega}^2}n_B^2 + \frac{\nu}{2\pi^2}\int_0^{k_F}k^2 d|\mathbf{k}|\sqrt{k^2 + m^{*2}},\tag{3.117}$$

$$P = -\frac{1}{2}\frac{m_{\sigma}^2}{g_{\sigma}^2}(m - m^*)^2 + \frac{1}{2}\frac{g_{\omega}^2}{m_{\omega}^2}n_B^2 + \frac{1}{3}\frac{\nu}{2\pi^2}\int_0^{k_F}d|\mathbf{k}|\frac{k^4}{\sqrt{k^2 + m^{*2}}}.\tag{3.118}$$

The first two contributions to the right-hand side of the above equations originate from the mass terms associated with the vector and scalar fields, while the remaining term gives the energy density and pressure of a relativistic Fermi gas of nucleons of mass m^*, given by Eq. (3.104)

$$m^* = m - \frac{g_\sigma^2}{m_\sigma^2} \frac{\nu}{2\pi^2} \int_0^{k_F} \mathbf{k}^2 d|\mathbf{k}| \frac{m^*}{\sqrt{|\mathbf{k}|^2 + m^{*2}}}$$

$$= m - \frac{g_\sigma^2}{m_\sigma^2} \frac{m^*}{\pi^2} \left[k_F e_F^* - m^{*2} \ln\left(\frac{k_F + e_F^*}{m^*}\right) \right], \qquad (3.119)$$

with $e_F^* = \sqrt{k_F^2 + m^{*2}}$. Equations (3.117)–(3.119) yield energy density and pressure of nuclear matter as a function of the baryon number density $n_B = \nu k_F^3/(6\pi^2)$. The values of the unknown coefficients (m_σ^2/g_σ^2) and (m_ω^2/g_ω^2) can be determined by a fit to the empirical saturation properties of nuclear matter, that is, requiring that

$$\frac{B}{A} = \frac{\epsilon(n_0)}{n_0} - m = -16 \text{ MeV}, \qquad (3.120)$$

with $n_0 = 0.16 \text{ fm}^{-3}$. This procedure leads to the result

$$\frac{g_\sigma^2}{m_\sigma^2} m^2 = 267.1 \quad , \quad \frac{g_\omega^2}{m_\omega^2} m^2 = 195.9 . \qquad (3.121)$$

The solid and dashed lines of Fig. 3.11 show the energies per nucleon of SNM and PNM predicted by the σ-ω model, plotted against the Fermi momentum k_F. Note that PNM is unbound at all densities.

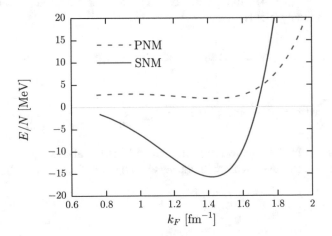

Fig. 3.11 Fermi momentum dependence of the energy per nucleon of SNM (solid line) and PNM (dashed line), evaluated using the σ-ω model in the mean field approximation, with the value of the parameters given by Eq. (3.121)

As a final remark, it has to be pointed out that the σ-ω model described in this section is the simplest implementation of the RMF formalism to nuclear matter. Following the pioneering work of Walecka [67], this approach has been extended by adding to the Lagrangian density the contributions of the vector-isovector ρ-meson, as well as non-linear self-interaction terms.

3.4   The Equation of State of Charge-Neutral β-Stable Matter

The calculation of neutron star properties requires the determination of the EOS of charge neutral matter consisting of neutrons, protons and leptons in equilibrium with respect to the neutron β-decay and lepton capture processes

$$n \to p + \ell + \bar{\nu}_\ell \quad , \quad p + \ell \to n + \nu_\ell \, , \tag{3.122}$$

where ℓ denotes the flavour of the lepton participating in the reactions; see also Chap. 2.

Let us consider matter comprising electrons and muons, hereafter referred to as $npe\mu$ matter. Under the assumption of transparency to neutrinos and antineutrinos, implying that these particles have vanishing chemical potentials, the equilibrium condition reduces to

$$\mu_n - \mu_p = \mu_e = \mu_\mu \, , \tag{3.123}$$

where μ_ℓ denotes the chemical potential of the leptons of flavour ℓ. The above condition must be fulfilled together with the additional constraint of charge neutrality, requiring that

$$x_p = x_e + x_\mu \, , \tag{3.124}$$

with $x_\ell = \varrho_\ell / \varrho_B$, ϱ_ℓ being the density of leptons of flavour ℓ. For any given value of the baryon number density n_B, Eqs. (3.123) and (3.124) uniquely determine the proton and lepton fractions, needed to obtain the EOS of β-stable matter.

The proton fraction of charge neutral β-stable $npe\mu$ matter obtained using the APR model of Ref. [63] (solid line) and the model of Ref. [68] (dashed line)—based on the RMF formalism—are compared in Fig. 3.12. It should be noted that the two approaches, while yielding similar results in the region of nuclear and subnuclear densities—in which they are both constrained by phenomenology—predict very different values of x_p at larger values of n_B. The kink at $n_B \lesssim 0.2$ fm^{-3} signals a transition from the normal phase of nuclear matter to a spin-isospin ordered phase associated with the onset of neutral pion condensation [62].

Figure 3.13 shows the energy per baryon of charge neutral β-stable $npe\mu$ matter obtained using the nuclear matter EOS of Akmal Pandharipande and Ravenhall [63] and treating the leptons as ultrarelativistic non interacting particles. The results corresponding to SNM and PNM are also displayed, for comparison.

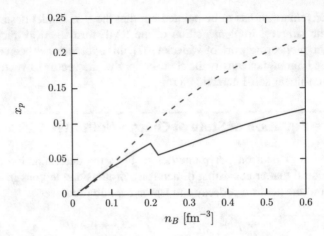

Fig. 3.12 Proton fraction of charge-neutral β-stable matter as a function of nucleon number density. The solid and dashed lines line represent the results of the APR model of Ref. [63] and the RMF calculations of Ref. [68], respectively

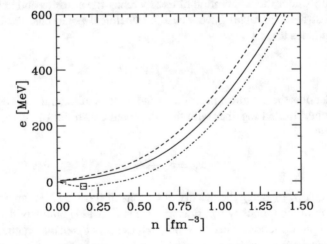

Fig. 3.13 Energy per baryon of nuclear matter calculated using the model of Ref. [63], plotted as a function of baryon number density. The dashed, dot-dash and solid lines correspond to PNM, SNM and charge-neutral β-stable matter, respectively. The box represents the empirical equilibrium properties of SNM obtained from extrapolation nuclear data

Appendix 1: Speed of Sound in Matter and Causality

According to the field-theoretical model proposed by Zel'dovich in the 1960s [36], matter consists of point like baryons interacting through exchange of a massive vector meson. The corresponding Lagrangian density is

$$\mathcal{L} = -\frac{1}{4}F_{\mu\nu}F^{\mu\nu} - \frac{1}{2}\mu_V^2 A_\mu A^\mu - g J_\mu A^\mu, \tag{3.125}$$

where $F^{\mu\nu} = \partial^\mu A^\nu - \partial^\nu A^\mu$, $A^\mu \equiv (\phi, \mathbf{A})$, μ_V is the meson mass and g denotes the coupling constant. The field equation obtained from the above $\mathcal{L}$ turns out to be

$$(\partial^\nu \partial_\nu + \mu_V^2) A_\mu = g J_\mu. \tag{3.126}$$

In the simple case of a pointlike stationary source located at $\mathbf{x} = 0$

$$J_\mu \equiv (J_0, \mathbf{J}) \equiv (\delta(\mathbf{x}), 0), \tag{3.127}$$

and the solution of Eq. (3.126) reads

$$\phi(\mathbf{x}') = g\,\frac{e^{-\mu|\mathbf{x}'-\mathbf{x}|}}{|\mathbf{x}'-\mathbf{x}|}, \quad \mathbf{A} = 0. \tag{3.128}$$

Within the above model, two charges at rest separated by a distance r repel each other with a force of magnitude

$$F = -g^2 \frac{d}{dr}\frac{e^{-\mu_V r}}{r}, \tag{3.129}$$

with the corresponding interaction energy being

$$g\phi = g^2 \frac{e^{-\mu_V r}}{r}. \tag{3.130}$$

In the case of N particles of mass M, since the equation of motion (3.126) is linear, one can use the superposition principle to write the total energy as

$$E = NM + g^2 \sum_{j>i=1}^{N} \frac{e^{-\mu_V r_{ij}}}{r_{ij}}, \tag{3.131}$$

with $r_{ij} = |\mathbf{r}_i - \mathbf{r}_j|$. Let us now make the further assumption that the average particle number density, n, is such that

$$\left(\frac{1}{n}\right)^{1/3} \ll \frac{1}{\mu_V}. \tag{3.132}$$

The above equation implies that the meson field varies slowly over distances comparable to the average particle separation. Under this condition one can use the mean field approximation, and rewrite Eq. (3.131) in the form

$$e = \frac{E}{N} = M + \frac{g^2}{2} \int d^3r \frac{e^{-\mu_V r}}{r} = m + 2\pi g^2 \frac{n}{\mu_V} . \tag{3.133}$$

The corresponding expressions of energy density and pressure read

$$\epsilon = ne = nM + 2\pi g^2 \frac{n^2}{\mu_V}, \tag{3.134}$$

and

$$P = n^2 \left(\frac{\partial e}{\partial n} \right) = 2\pi g^2 \frac{n^2}{\mu_V} . \tag{3.135}$$

From the above equations it follows that in the $n \to \infty$ limit $P \to \epsilon$, implying in turn $c_s^2 \to 1$.

In conclusion, Zel'dovich work shows that, even within the framework of a relativistically consistent formalism, the causality limit—corresponding to $\epsilon \propto n^2$—is indeed attained assuming a simple semirealistic model of nuclear matter, in which nucleons interact through exchange of a vector meson.

Appendix 2: Derivation of Yukawa's OPE Potential

The Two-Nucleon System

Neglecting a $\sim 1\%_0$ mass difference, proton and neutron can be viewed as two states of the same spin $1/2$ particle, the nucleon (N), specified by an additional quantum number dubbed isospin.

In the absence of interactions, the nucleon is described by the equation of motion derived from the Lagrangian density

$$\mathcal{L}_0 = \bar{\psi}_N \left(i\gamma^\mu \partial_\mu - m \right) \psi_N . \tag{3.136}$$

where the γ^μ are Dirac's matrices satisfying $\{\gamma^\mu, \gamma^\mu\} = 2g^{\mu\nu}$, $g^{\mu\nu}$ being the metric tensor of Minkowski space, and the field ψ_N can be conveniently written in the form

$$\psi_N = \begin{pmatrix} \psi_p \\ \psi_n \end{pmatrix} , \tag{3.137}$$

with ψ_p and ψ_n being the spinor fields associated with the proton and the neutron, respectively.

The Lagrangian density of Eq. (3.136) is invariant under the SU(2) global phase transformation

$$U = e^{i\alpha_j \tau_j} , \tag{3.138}$$

where the α_j are constants, the τ_j are Pauli matrices acting in isospin space and a sum over the index j is understood. Proton and neutron correspond to isospin projections $+1/2$ and $-1/2$, respectively.

Proton-proton and neutron-neutron pairs always have total isospin T=1 whereas a proton-neutron pair may have either $T = 0$ or $T = 1$. The two-nucleon isospin states $|T, T_3\rangle$ can be can be written in terms of proton and neutron degrees of freedom as

$$|0, 0\rangle = \frac{1}{\sqrt{2}} (|pn\rangle - |np\rangle) , \tag{3.139}$$

$$|1, 1\rangle = |pp\rangle \quad , \quad |1, 0\rangle = \frac{1}{\sqrt{2}} (|pn\rangle + |np\rangle) \quad , \quad |1, -1\rangle = |nn\rangle .$$

Isospin invariance implies that the interaction between two nucleons coupled to total spin S depends on their total isospin T, but *not* on its projection T_3. For example, the force acting between two protons with total spin $S = 0$ is the same as that acting between a proton and a neutron with spins and isospins coupled to $S = 0$ and $T = 1$.

Besides the observed properties of the two-nucleon bound state—the nucleus ^{2}H, or deuteron—the large data base of phase shifts accurately measured by NN scattering experiments, comprising $\sim$4000 data points at beam energies up to 350 MeV in the lab frame, provides valuable additional information on the nature of NN interactions.

The Two-Nucleon Interaction

The theoretical description of the NN interaction within the framework of quantum field theory was first proposed by Yukawa in in the 1930s [37]. He made the hypothesis that nucleons interact through the exchange of a particle whose mass, μ, is related to the range of the interaction, r_0, through $r_0 \sim 1/\mu$.[2] Using $r_0 \sim 1$ fm $= 10^{-13}$ cm, this relation yields $\mu \sim 200$ MeV.

Yukawa's idea has been successfully implemented identifying the exchanged particle with the π meson (or *pion*), discovered in 1947, the mass of which is $m_\pi \approx 140$ MeV. Experimental data show that the pion is a pseudoscalar, i.e. spin-parity 0^-, particle, occurring in three different charge states denoted π^0, π^+, and π^-. As a consequence, it can be regarded as an isospin triplet having $T = 1$, with the charge states being associated with the projections $T_3 = 0, \pm 1$.

---

[2] We adopt the system of units in which $\hbar = c = 1$, implying in turn that 1 fm^{-1} = 197.3 MeV.

The simplest π-nucleon interaction Lagrangian compatible with the requirement of Lorentz invariance and with the observation that nuclear interactions conserve parity involves a *pseudoscalar* coupling, and can be written in the form [38]

$$\mathcal{L}_I = -ig\bar{\psi}_N \gamma^5 \tau^j \psi_N \pi^j , \qquad (3.140)$$

where

$$\pi^1 = \frac{1}{\sqrt{2}}(\pi^+ + \pi^-) \ , \quad \pi^2 = \frac{i}{\sqrt{2}}(\pi^+ - \pi^-) \ , \quad \pi^3 = \pi^0 . \qquad (3.141)$$

and a sum over the index j is understood.

In Eq. (3.140), g is the pseudoscalar strong interactions coupling constant and $\gamma^5 = i\gamma^0\gamma^1\gamma^2\gamma^3$. The γ^μ are again Dirac matrices, and the Pauli matrices τ^j, acting in isospin space, are associated with the isospin of the nucleon. Note that, in the non relativistic limit, the pseudoscalar coupling, $-ig\gamma^5\tau$, and the alternative pseudovector coupling, $ig'\gamma^5\gamma^\mu\tau\partial_\mu$, yield the same NN potential. It is apparent that in both cases the isospin formalism allows to take into account in a concise fashion all interaction vertices, involving proton, neutrons, charged pions and neutral pions.

Let us consider the NN scattering process

$$N(p_1 s_1) + N(p_2 s_2) \rightarrow N(p'_1 s'_1) + N(p'_2 s'_2),$$

depicted by the Feynman diagrams of Fig. 3.14.

The corresponding S-matrix element reads

$$S_{fi} = (-ig)^2 \frac{m^2}{\left(E_1 E_2 E'_1 E'_2\right)^{1/2}} (2\pi)^4 \delta^{(4)}(p_1 + p_2 - p'_1 - p'_2) \qquad (3.142)$$

$$\times \left\{ \left[\eta^\dagger_{1'} \bar{u}_{1'} i\gamma_5 \tau u_1 \eta_1 \right] \frac{i}{k^2 - m_\pi^2} \left[\eta^\dagger_{2'} \bar{u}_{2'} i\gamma_5 \tau u_2 \eta_2 \right] - \left[(1', 2') \rightleftharpoons (2', 1') \right] \right\} ,$$

where m_π is the pion mass, $k = p_1 - p'_1 = p'_2 - p_2$, and η_i denotes the two-component Pauli spinor describing the isospin state of particle i.

The direct term of Eq. (3.142) can be rewritten in the form

$$S_{fi}^{(D)} = ig^2 \frac{m^2}{\left(E_1 E_2 E'_1 E'_2\right)^{1/2}} (2\pi)^4 \delta^{(4)}(p_1 + p_2 - p'_1 - p'_2) \qquad (3.143)$$

$$\times \frac{1}{k^2 - m_\pi^2} \times \eta^\dagger_{2'} \tau \eta_2 \ \bar{u}_{2'} \gamma_5 u_2 \ \bar{u}_{1'} \gamma_5 u_1 \ \eta^\dagger_{1'} \tau \eta_1 .$$

Substituting the expression of the Dirac spinor describing a particle with momentum $\mathbf{p}$, energy $E = \sqrt{\mathbf{p}^2 + m^2}$, and spin projection s

$$u_s(\mathbf{p}) = \sqrt{\frac{E+m}{2E}} \begin{pmatrix} \chi_s \\ \frac{\sigma \cdot \mathbf{p}}{E+m} \chi_s \end{pmatrix} , \qquad (3.144)$$

Fig. 3.14 Feynman diagrams representing the direct (upper panel) and exchange (lower panel) contributions to one-pion-exchange between two nucleons. The corresponding amplitude is given by Eq. (3.142)

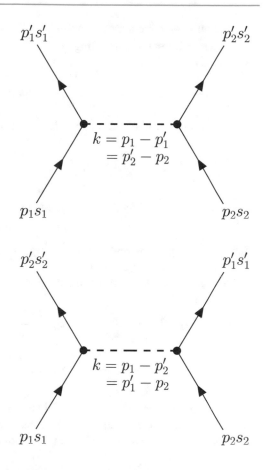

where χ_s is a Pauli spinor acting in spin space, and taking the non relativistic limit we find

$$\bar{u}_{2'}\gamma_5 u_2 = \sqrt{\frac{(E'_2 + m)(E_2 + m)}{4E'_2 E_2}}$$

$$\times \left(\chi^\dagger_{2'} \frac{\sigma \cdot \mathbf{p}_2}{E_2 + m} \chi_2 - \chi^\dagger_{2'} \frac{\sigma \cdot \mathbf{p}'_2}{E'_2 + m} \chi_2 \right) \tag{3.145}$$

$$\approx \chi^\dagger_{2'} \frac{\sigma (\mathbf{p}_2 - \mathbf{p}'_2)}{2m} \chi_2 = -\chi^\dagger_{2'} \frac{(\sigma \cdot \mathbf{k})}{2m} \chi_2 \,,$$

and the similar expression for $\bar{u}_{1'}\gamma_5 u_1$.

The non relativistic approximation also allows to use $E_i \approx E_i' \approx m$ and $k^2 = (E_i - E_i')^2 - \mathbf{k}^2 \approx -\mathbf{k}^2$, implying

$$\frac{1}{k^2 - m_\pi^2} \approx \frac{-1}{\mathbf{k}^2 + m_\pi^2} , \qquad (3.146)$$

and substituting the above results in the definition of the amplitude S_{fi} we obtain

$$S_{fi}^{(D)} \approx -i \frac{g^2}{4m^2} (2\pi)^4 \delta^{(4)}(p_1 + p_2 - p_1' - p_2') \qquad (3.147)$$

$$\times \langle f| \left[(\boldsymbol{\tau}_1 \cdot \boldsymbol{\tau}_2) \frac{(\boldsymbol{\sigma}_1 \cdot \mathbf{k})(\boldsymbol{\sigma}_2 \cdot \mathbf{k})}{\mathbf{k}^2 + m_\pi^2} \right] |i\rangle ,$$

where $|i\rangle = \eta_1 \eta_2 \, \chi_1 \chi_2$. and $\langle f| = \eta_{1'}^\dagger \eta_{2'}^\dagger \, \chi_{1'}^\dagger \chi_{2'}^\dagger$ denote the initial and final states of the interacting particles, respectively.

Equation (3.147) suggest that the operator

$$v_{\mathrm{OPE}}(\mathbf{k}) = -\frac{g^2}{4m^2} (\boldsymbol{\tau}_1 \cdot \boldsymbol{\tau}_2) \frac{(\boldsymbol{\sigma}_1 \cdot \mathbf{k})(\boldsymbol{\sigma}_2 \cdot \mathbf{k})}{|\mathbf{k}|^2 + m_\pi^2}$$

$$= -\left(\frac{f}{m_\pi}\right)^2 (\boldsymbol{\tau}_1 \cdot \boldsymbol{\tau}_2) \frac{(\boldsymbol{\sigma}_1 \cdot \mathbf{k})(\boldsymbol{\sigma}_2 \cdot \mathbf{k})}{|\mathbf{k}|^2 + m_\pi^2} , \qquad (3.148)$$

can be interpreted as the one-pion-exchange potential in momentum space. Note that in the second line of Eq. (3.148) we have replaced the pseudoscalar coupling constant g with the new dimensionless constant (use $g^2/4\pi \approx 14$)

$$f^2 = g^2 \frac{m_\pi^2}{4m^2} \approx 4\pi \times 14 \frac{(140)^2}{4 \times (939)^2} \approx 4\pi \times 0.08 \approx 1 . \qquad (3.149)$$

The coordinate-space potential is obtained from Fourier transformation, leading to

$$v_{\mathrm{OPE}}(\mathbf{r}) = -\left(\frac{f}{m_\pi}\right) (\boldsymbol{\tau}_1 \cdot \boldsymbol{\tau}_2) (\boldsymbol{\sigma}_1 \cdot \boldsymbol{\nabla}) (\boldsymbol{\sigma}_2 \cdot \boldsymbol{\nabla}) \int \frac{d^3k}{(2\pi)^3} \frac{1}{(\mathbf{k}^2 + m_\pi^2)} e^{-i\mathbf{k}\cdot\mathbf{r}} , \qquad (3.150)$$

where

$$\int \frac{d^3k}{(2\pi)^3} \frac{1}{(\mathbf{k}^2 + m_\pi^2)} e^{-i\mathbf{k}\cdot\mathbf{r}} = \frac{1}{4\pi} \frac{e^{-m_\pi r}}{r} = \frac{1}{4\pi} y_\pi(r) .$$

The gradients appearing in Eq. (3.150) can be readily evaluated exploiting the relation

$$(-\nabla^2 + m_\pi^2)\, y_\pi(r) = 4\pi \delta(\mathbf{r})\,, \tag{3.151}$$

to rewrite

$$(\boldsymbol{\sigma}_1 \cdot \boldsymbol{\nabla})(\boldsymbol{\sigma}_2 \cdot \boldsymbol{\nabla}) y_\pi(r) \tag{3.152}$$

$$= \left[(\boldsymbol{\sigma}_1 \cdot \boldsymbol{\nabla})(\boldsymbol{\sigma}_2 \cdot \boldsymbol{\nabla}) - \frac{1}{3}(\boldsymbol{\sigma}_1 \cdot \boldsymbol{\sigma}_2)\nabla^2\right] y_\pi(r) + \frac{1}{3}(\boldsymbol{\sigma}_1 \cdot \boldsymbol{\sigma}_2)\nabla^2\, y_\pi(r)\,.$$

The δ-function contribution to $\nabla^2 y_\pi(r)$, arising from Eq. (3.151), does not appear in the first term, yielding

$$\left[(\boldsymbol{\sigma}_1 \cdot \boldsymbol{\nabla})(\boldsymbol{\sigma}_2 \cdot \boldsymbol{\nabla}) - \frac{1}{3}(\boldsymbol{\sigma}_1 \cdot \boldsymbol{\sigma}_2)\nabla^2\right] y_\pi(r) \tag{3.153}$$

$$= \left[(\boldsymbol{\sigma}_1 \cdot \hat{\mathbf{r}})(\boldsymbol{\sigma}_2 \cdot \hat{\mathbf{r}}) - \frac{1}{3}(\boldsymbol{\sigma}_1 \cdot \boldsymbol{\sigma}_2)\right]\left(m_\pi^2 + \frac{3m_\pi}{r} + \frac{3}{r^2}\right) y_\pi(r),$$

where $\hat{\mathbf{r}} = \mathbf{r}/|\mathbf{r}|$. In the second term, it can be replaced with $m_\pi^2\, y_\pi(r) - 4\pi\,\delta(\mathbf{r})$ using Eq. (3.151).

Performing the calculation of the derivatives in Eq. (3.150) finally leads to

$$V_{\text{OPE}}(\mathbf{r}) = \frac{1}{3}\frac{1}{4\pi}\, f^2\, m_\pi\, (\boldsymbol{\tau}_1 \cdot \boldsymbol{\tau}_2)\left[T_\pi(r)S_{12}\left(Y_\pi(r) - \frac{4\pi}{m_\pi^3}\delta(\mathbf{r})\right)(\boldsymbol{\sigma}_1 \cdot \boldsymbol{\sigma}_2)\right], \tag{3.154}$$

with

$$Y_\pi(r) = \frac{e^{-m_\pi r}}{m_\pi r}\,, \tag{3.155}$$

and

$$T_\pi(r) = \left(1 + \frac{3}{m_\pi r} + \frac{3}{m_\pi^2 r^2}\right) Y_\pi(r)\,. \tag{3.156}$$

Note that due to the presence of a contribution involving the operator

$$S_{12} = \frac{3}{r^2}(\boldsymbol{\sigma}_1 \cdot \mathbf{r})(\boldsymbol{\sigma}_2 \cdot \mathbf{r}) - (\boldsymbol{\sigma}_1 \cdot \boldsymbol{\sigma}_2)\,, \tag{3.157}$$

reminiscent of the operator describing the interaction between two magnetic dipoles, the above potential is *not* spherically symmetric.

Yukawa's potential provides a good description of the long range part ($|\mathbf{r}| > 1.5$ fm) of the NN interaction, as shown by the fit to the NN scattering phase shifts in states of high angular momentum. In these states, the probability of finding the two nucleons at small relative distances is, in fact, largely suppressed by the presence of a strong centrifugal barrier.

Exotic Forms of Matter

4

Abstract

The density of the neutron star core is believed to exceed the central density of atomic nuclei by as much as a factor five. At these extreme densities, the stable configuration of matter is expected to involve *exotic* constituents, other that neutrons, protons and leptons. This chapter provides an overview of the mechanisms leading to the appearance of strange baryons and deconfined quarks in dense nuclear matter. The theoretical models most commonly employed for the description of the exotic phases, and the nature of the transition from nuclear matter to quark matter is also discussed.

4.1 Stability of Strange Baryonic Matter

As the density grows up to values well beyond nuclear saturation density, the appearance of *exotic* forms of matter, comprising baryons other than protons and neutrons, can become energetically favoured. Strange baryons, or hyperons, are produced through weak interaction processes, such as

$$p + e \to \Lambda^0 + \nu_e , \tag{4.1}$$

$$p + e \to \Sigma^0 + \nu_e, \tag{4.2}$$

$$n + e \to \Sigma^- + \nu_e. \tag{4.3}$$

For example, the reaction (4.3), leading to the appearance of a Σ^-, sets in as soon as the sum of the proton and electron chemical potentials exceeds the rest mass of the produced baryon, M_{Σ^-}. Here, we assume that neutrinos do not interact significantly with matter, and therefore they have vanishing density and chemical potentials. The results of theoretical calculations suggest that in cold neutron stars the threshold

© The Author(s), under exclusive license to Springer Nature Switzerland AG 2023
O. Benhar, *Structure and Dynamics of Compact Stars*, Lecture Notes
in Physics 1019, https://doi.org/10.1007/978-3-031-35628-5_4

Table 4.1 Hyperon properties. J and I_3 denote the spin and the isospin projection, respectively

	Charge	Mass (MeV)	J	I_3	Valence quark structure	Strangeness
Λ^0	0	1115.7	1/2	0	uds	-1
Σ^-	-1	1197.4	1/2	-1	dds	-1
Σ^0	0	1192.6	1/2	0	uds	-1
Σ^+	$+1$	1189.4	1/2	$+1$	uus	-1
Ξ^0	0	1314.9	1/2	$+1/2$	uss	-2
Ξ^-	-1	1321.7	1/2	$-1/2$	dss	-2

condition is typically fulfilled at baryon number densities $n_B \gtrsim 2n_0$, with n_0 being the equilibrium density of SNM, and that at $n_B \approx 3n_0$ hyperon may account for a substantial fraction of the total baryon population. The properties of baryons with non-zero strangeness, or hyperons, are summarised in Table 4.1.

The mechanism responsible for the stability of strange hadronic matter—driven by the Fermi-Dirac statistics obeyed by the constituent particles—is similar to that leading to neutronisation, discussed in Chap. 2.

Let us consider a system consisting of B baryonic species $b_1 \ldots b_B$ and L leptonic species $\ell_1 \ldots \ell_L$, in equilibrium with respect to the weak interaction processes

$$b_i \rightarrow b_j + \ell_k + \bar{\nu}_\ell \,, \tag{4.4}$$

$$b_j + \ell_k \rightarrow b_i + \nu_\ell \,, \tag{4.5}$$

with $i, j = 1 \ldots B$ and $k = 1 \ldots L$. The ground state of system, specified by the number densities of the constituent particles, $\{n_{b_i}\}$ and $\{n_{\ell_i}\}$, is determined through minimisation of the energy density, with the constraints dictated from conservation of the baryon density, n_B, and charge neutrality, implying

$$\sum_{i=1}^{B} n_{b_i} = n_B \,, \tag{4.6}$$

$$\sum_{i=1}^{B} Q_{b_i} n_{b_i} + \sum_{i=1}^{L} Q_{\ell_i} n_{\ell_i} = 0 \,, \tag{4.7}$$

where Q_{b_i} and Q_{ℓ_i} denote the electric charge of the i-th baryonic and leptonic species, respectively.

To see how the determination of matter composition works, consider, as an example, a system consisting of protons (p), neutrons (n), electrons (e), muons (μ), and the hyperons Σ^- and Λ^0. In addition to Eqs. (4.6) and (4.7), which in this case take the form

$$n_n + n_p + n_\Lambda + n_\Sigma = n_B \,,$$

and

$$n_p - n_e - n_\mu - n_\Sigma = 0 ,$$

we obtain $B + L - 2 = 4$ equilibrium equations involving the chemical potentials

$$\mu_p = \mu_n - \mu_e ,$$

$$\mu_{\Sigma^-} = \mu_n + \mu_e ,$$

$$\mu_{\Lambda^0} = \mu_n ,$$

$$\mu_\mu = \mu_e .$$

It follows that there are, in fact, only two independent chemical potentials, e.g. μ_n and μ_p.

In principle, for any given baryon density, n_B, the densities of the constituent particles can be determined from the above equations. Treating all constituents as non interacting particles, one finds that Σ^- and Λ^0 appear at densities $\sim 4n_0$ and $\sim 8n_0$, respectively, n_0 being the equilibrium density of isospin-symmetric nuclear matter; see Refs. [69, 70].

It should be kept in mind, however, that the density at which production of a strange hadron is expected to set in does not depend on its mass only. To see this, consider the Λ^0 and Σ^- hyperons, whose appearance becomes energetically favoured as soon as the threshold conditions

$$\mu_n + \mu_e = M_{\Sigma^-} ,$$

$$\mu_n = M_{\Lambda^0} ,$$

are fulfilled. From the above relations it follows that, in spite of the larger mass, the threshold density of Σ^- production is in fact lower than that of Λ^0 production if the electron chemical potential is such that

$$\mu_e > M_{\Sigma^-} - M_{\Lambda^0} \approx 80 \text{ MeV} .$$

Treating the electrons as a relativistic Fermi gas, the corresponding chemical potential can be easily obtained, and the threshold condition can be written in the form

$$\mu_e = \sqrt{p_{F_e}^2 + m_e^2} \approx p_{F_e} = (3\pi^2 n_e)^{1/3} > 80 \text{ MeV} ,$$

implying

$$n_e \gtrsim 2 \times 10^{-3} \text{ fm}^{-3} .$$

Under the reasonable assumption that the electron density be of the order of one percent of the baryon density, the estimated value of n_e corresponds to $n_B >$ 0.2 fm$^{-3} \gtrsim n_0$, a density certainly reached in compact stars.

The above discussion suggests that, as density increases, the Σ^- is the first hyperon to appear, followed closely by the Λ^0. However, the formation of Σ^- is expected to be quickly quenched by isospin-dependent forces, that tend to disfavour an excess of Σ^- over Σ^+ as well as a joint excess of Σ^- and neutrons, both having negative isospin projection.

4.1.1 Hyperon Interactions

Unlike lepton interactions, that are always negligible, the interactions of baryons, including hyperons, have a critical impact on matter composition, which in turn determines the EOS. The appearance of strange baryons has important implications for the global properties of neutron stars, such as mass and radius, that will be discussed in Chap. 7.

In principle, the EOS of strange baryonic matter can be studied using either non relativistic many-body theory or relativistic mean field theory, as discussed in Sects. 3.3.4 and 3.3.7, respectively. However, the former approach is based on potentials describing hyperon-nucleon (YN) and hyperon-hyperon (YY) interactions, while the latter involves the masses of the exchanged bosons, as well as the YN and YY coupling constants. In either case, the determination of a dynamical model requires the availability of experimental input.

The data providing information on YN and YY forces are scarce and not very accurate. Owing to the short lifetime of hyperons and to the low beam density, YN scattering experiments entail severe problems, and the available set of YN scattering phase shifts comprises only ~50 data points, to be compared to the ~5000 phase shifts precisely measured in NN scattering. No data at all is available in the YY sector.

Complementary information is provided by the measured properties of hypernuclei.[1] Systematic studies, carried out using different experimental techniques, have allowed the determination of the energy levels of the Λ_0 single-particle major shells in a variety of hypernuclei, having mass in the range $7 \leq A \leq 208$. The results of these analyses are illustrated in Fig. 4.1.

A combination of hypernuclear and scattering data suggests that the Λp interaction is weaker than either the pn or pp interactions. Some further insight on the relation between strong interactions in the nucleon and hyperon sectors can be obtained from an admittedly crude argument based on the quark model of

---

[1] Hypernuclei are similar to a conventional atomic nuclei, but contain at least one hyperon in addition to protons and neutrons. For example, the hypernucleus $^{12}C_\Lambda$ consists of 12 baryons, one of which is a Λ.

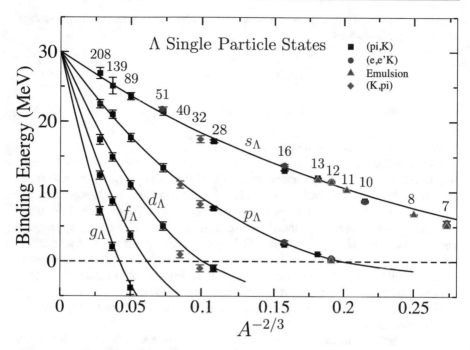

Fig. 4.1 Energy levels of the Λ_0 single-particle major shells in hypernuclei having mass in the range $7 \leq A \leq 208$. Reprinted from Ref. [71] with permissions, © APS 2016. All rights reserved

hadrons [72]. According to this picture, the π-meson is a quark-antiquark system, with the composition of the three charge states being

$$\pi^+ = u\bar{d} \quad , \quad \pi^- = d\bar{u} \quad , \quad \pi^0 = \frac{1}{\sqrt{2}}(u\bar{u} - d\bar{d}) \, ,$$

where u and d denote the up and down quark, having charge $2/3$ and $-1/3$, respectively. As discussed in Chap. 3, the intermediate-range NN force can be described in terms of exchange of a quark-antiquark system modelled by a scalar-isoscalar meson, whose structure can be written in the form

$$\sigma = \frac{1}{\sqrt{2}}(u\bar{u} + d\bar{d}) \, .$$

Note that the right-hand side of the above equation is essentially the operator counting the number of non-strange quarks. Because nucleons contain three non-strange quarks and the Λ and Σ only contain two, the coupling constants of the Λp or Σp interaction mediated by the σ-meson is $2/3$ of the corresponding pp coupling constant. In the language of nuclear many-body theory, this argument leads to predict the following simple relations between interaction potentials in the strange and non-strange sectors

$$v_{\Lambda N} = v_{\Sigma N} = \frac{2}{3} v_{NN} \,,$$

$$v_{\Lambda\Lambda} = v_{\Lambda\Sigma} = v_{\Sigma\Sigma} = \frac{4}{9} v_{NN} \,,$$

where N denotes either a proton or a neutron.

More realistic and accurate descriptions of the YN interaction based on the boson exchange model, referred to as Jülich [73] and Nijmegen [74] potentials, have been developed exploiting all the available information on YN scattering and hypernuclear properties. The results of microscopic calculations carried out using these potentials suggest that the threshold densities for Σ^- and Λ_0 appearance are in the range $\sim(1.5 - 2)n_0$ and $\sim(2.5 - 4)n_0$, respectively [75, 76]. In recent years, hyperon interaction potentials have also been obtained within the framework of chiral effective field theory; see, e.g., Ref. [77]. Figure 4.2 illustrates the typical composition of matter resulting from a calculation carried out using the formalism of nuclear many-body theory, a realistic nuclear Hamiltonian and the Nijmegen model of YN interactions.

All theoretical calculations indicate that the appearance of hyperons leads to a significant softening of the EOS, which severely hampers the capability to support stable stars with mass largely exceeding the canonical value of $\sim 1.4 \, M_\odot$; see Fig. 4.3. The difficulty to reconcile the appearance of strange baryons with the observations of neutron stars having masses $\gtrsim 2 \, M_\odot$ is often referred to as *hyperon puzzle*.

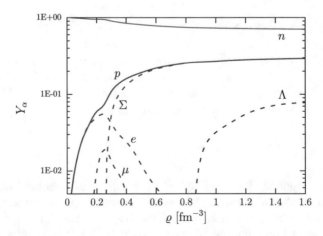

Fig. 4.2 Composition of strange baryonic matter obtained from non relativistic nuclear many-body theory using the Nijmegen model of YN interactions. Adapted from Ref. [78] with permissions, © APS 2006. All rights reserved

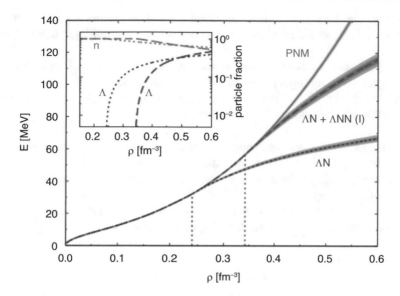

Fig. 4.3 Equations of state of PNM and baryonic matter including Λs. The curve labelled $\Lambda N + \Lambda NN$ represent results obtained taking into account the effects of three-body ΛNN interactions. The dashed and dotted lines in the inset illustrate the dependence of matter composition on the ΛNN potential model. Reprinted from Ref. [79] with permissions, © APS 2015. All rights reserved

4.2   Deconfinement and Quark Matter

An ab initio description of the QCD phase diagram requires the use of non-perturbative techniques, notably the numerical implementations of quantum field theory on a lattice; see, e.g., [80, 81]. However, the application of this approach in the region corresponding to non vanishing chemical potential involves serious difficulties. As a consequence, most studies of the properties of quark matter are carried out within simplified QCD-inspired models, involving a set of parameters whose values are determined in such a way as to reproduce the observed properties of hadrons.

4.2.1   The MIT Bag Model

This section is devoted to the description of a model developed in the 1970s at the Massachusetts Institute of Technology (MIT)—referred to as *bag model*—in which the main features of QCD are taken into account in a crude and yet very effective fashion. It is remarkable that, in spite of its simplicity, this model provides a reasonable description of the hadronic spectrum.

Within the MIT bag model, the fundamental properties of QCD, that is, *asymptotic freedom* and *confinement*, translate into two simple hypotheses:

- the QCD vacuum exerts a pressure on any quark aggregate of vanishing color charge, thus confining it to a finite region in space: the bag;
- within the bag, gluon-exchange interactions between quarks are negligible, or tractable at lowest order of perturbation theory.

In the simplest implementation of the above picture, in which the quarks within the bag are described as non-interacting particles obeying Fermi statistics, the thermodynamic potential can be computed explicitly using the formalism of quantum field theory.

Let us consider a system consisting of N_f types of fermions of masses m_i, with $i = 1, \ldots, N_f$, described by the Dirac Lagrangian density

$$\mathcal{L} = \sum_{i=1}^{N_f} \bar{q}^i (i\slashed{\partial} - m_i)q^i \, , \tag{4.8}$$

where q^i denotes the fermion field. The formalism suitable to describe fields satisfying anticommutation rules, is based on the use of N_f Grassmann variables η_i, whose gaussian integral is given by

$$\int d\eta_1^\dagger d\eta_1 \cdot d\eta_N^\dagger d\eta_N e^{\eta_i^\dagger A^{ij}\eta_j} = \det(A) \, . \tag{4.9}$$

Using the techniques outlined in the Appendix, we can write the partition function of the system in the form [82]

$$Z = \prod_i \int i\mathcal{D}q^{i\dagger} \int \mathcal{D}q^i \tag{4.10}$$

$$\times \exp\left[\int_0^\beta d\tau \int d^3x \, \bar{q}^i \left(-\gamma^0 \partial_\tau + i\boldsymbol{\gamma} \cdot \boldsymbol{\nabla} - m_i + \mu_i \gamma^0 \right) q^i \right] ,$$

where $\beta = 1/T$, and the functional integration is extended to all paths satisfying the antiperiodic boundary conditions $q^i(\mathbf{x}, 0) = -q^i(\mathbf{x}, \beta)$.

It is convenient to transform to momentum space using the expansion

$$q_\alpha^i(\mathbf{x}, \tau) = \frac{1}{\sqrt{V}} \sum_{n,\mathbf{p}} e^{i(\mathbf{p}\cdot\mathbf{x} + E_{in}\tau)} \tilde{q}_{\alpha n}^i(\mathbf{p}) \, , \tag{4.11}$$

where α is the index labelling the components of Dirac's spinors, and antiperiodicity requires that

$$E_{in} = (2n + 1)\pi T .   \tag{4.12}$$

From the above equations, it follows that the partition function can be written in the form

$$Z = \prod_{i,\alpha,\mathbf{p},n} \int i\mathcal{D}\,\tilde{q}^{i\dagger}_{\alpha n} \int \mathcal{D}\,\tilde{q}^{i}_{\alpha n}\ \exp i \sum_{in\mathbf{p}} \tilde{q}^{i\dagger}_{\alpha n} D^{\alpha\rho} \tilde{q}^{i}_{\rho n} = \det D,   \tag{4.13}$$

with

$$D = -i\beta\left[(-iE_{in} + \mu_i) - \gamma^0\boldsymbol{\gamma} \cdot \mathbf{p} - m_i\gamma^0\right].   \tag{4.14}$$

In addition, exploiting the operator relation

$$\ln[\det(D)] = \mathrm{Tr}\left[\ln(D)\right],   \tag{4.15}$$

one can derive from Eq. (4.14) the thermodynamic potential, defined as

$$\Omega = -T \ln Z = -PV .   \tag{4.16}$$

The results is

$$\begin{aligned}
\Omega &= -T \ln[\det(D)] \\
&= -2T \sum_{i,n,\mathbf{p}} \ln\left\{\beta^2[(E_{in} + \mu_i)^2 + E_{ip}^2]\right\} \\
&= -T \sum_{i,n,\mathbf{p}} \ln\left\{\beta^2[E_{in}^2 + (E_{ip} - \mu_i)^2]\right\} + \ln\left\{\beta^2[E_{in}^2 + (E_{ip} + \mu_i)^2]\right\} ,
\end{aligned}   \tag{4.17}$$

where $E_{ip} = \sqrt{\mathbf{p}^2 + m_i^2}$. Using Eq. (4.12) the logarithms appearing on the right-hand side of Eq. (4.17) can be rewritten in the form

$$\begin{aligned}
\ln&\left[(2n + 1)^2\pi^2 + \beta^2(E_{ip} \pm \mu_i)^2\right] \\
&= \int_1^{\beta^2(E_{ip}\pm\mu_i)^2} \frac{d\theta^2}{\theta^2 + (2n + 1)^2\pi^2} + \ln[1 + (2n + 1)^2\pi^2] ,
\end{aligned}   \tag{4.18}$$

which allows us to perform the sum over n using

$$\sum_{n=-\infty}^{\infty} \frac{1}{\theta^2 + (2n+1)^2 \pi^2} = \frac{1}{\theta}\left(\frac{1}{2} - \frac{1}{e^{\theta}+1}\right). \tag{4.19}$$

Integration over the variable θ yields an expression of the logarithm of the partition function that involves terms independent of β and the chemical potentials μ_i. These contributions do not affect the thermodynamics, because, after being exponentiated, can be included in the normalisation of the partition function. Neglecting them altogether leads to

$$\Omega = -6\,TV \sum_{i=1}^{N_f} \int \frac{d^3 p}{(2\pi)^3} [\beta E_{ip} + \ln(1 + e^{-\beta(E_{ip}-\mu_i)}) + \ln(1 + e^{-\beta(E_{ip}+\mu_i)})] , \tag{4.20}$$

where the factor $6 = 3 \times 2$ accounts for the degeneracy associated with colour and spin degrees of freedom. The second and third terms of the right-hand side represent the contributions arising from particles and antiparticles, respectively, while the first term contributes to the vacuum energy.

From the relation between pressure, thermodynamic potential and partition function, Eq. (4.79) of Appendix 1, one finally obtains

$$P_0 = 6\,T \sum_{i} \int \frac{d^3 p}{(2\pi)^3} \left\{ \ln\left[1 + \exp\beta(E_{ip} - \mu_i)\right] + \ln\left[1 + \exp\beta(E_{ip} + \mu_i)\right] \right\}. \tag{4.21}$$

In the bag model, one has to add to the pressure P_0, corresponding to a gas of non interacting quarks and antiquarks, three additional contributions: δP_{glue}, describing the pressure of a gluon gas, which can be computed using a technique similar to that employed to obtain Eq. (4.21), δP_{pert}, taking into account perturbative corrections arising from gluon exchange among quarks, and a constant term $-B$, parametrising the pressure exerted by the QCD vacuum on the perturbative vacuum within the bag. The final result is

$$P = P_0 + \delta P_{\text{glue}} + \delta P_{\text{pert}} - B , \tag{4.22}$$

with

$$\delta P_{\text{glue}} = -16 \int \frac{d^3 p}{(2\pi)^3} \ln[1 - e^{-\beta p}] . \tag{4.23}$$

Let us now consider the simplest implementation of the model, in which perturbative gluon exchange is disregarded and the number of flavours is limited to two, u and d, with $m_u = m_d = 0$ and $\mu_u = \mu_d = \mu$. Under these simplifying assumptions, the calculation of the pressure can be performed analytically, with the result [83]

$$P = -B + 37 \frac{\pi^2}{90} T^4 + \mu^2 T^2 + \frac{\mu^4}{2\pi^2} . \tag{4.24}$$

The factor 37 in front of the term proportional to T^4 results from the sum $16 + 21$, where $16 = 8 \times 2$ is the number of gluonic degrees of freedom and 21 is obtained by multiplying the number of degrees of freedom associated with quarks and antiquarks ($24 = 2 \times 3 \times 2 \times 2$) by the factor 7/8 characteristic of Fermi statistics.

Equation (4.24) can be employed to obtain a qualitative description of the QCD phase diagram. Assuming that a pion gas with vanishing chemical potential provides a schematic representation of the hadronic phase, we can write the corresponding pressure in the form [82]

$$P_{\text{had}} = 3 \frac{\pi^2}{90} T^4 , \tag{4.25}$$

where 3 is the number of pionic degrees of freedom, that is, the number of isospin states. Gibbs equilibrium condition is represented by the curve of Fig. 4.4, computed using the value $B = 57.5 \, \text{MeV fm}^{-3}$, determined from a fit to the masses and magnetic moments of light hadrons [84].

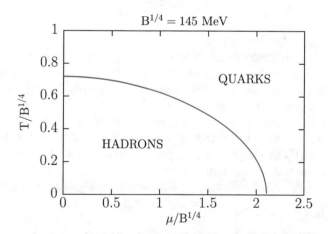

Fig. 4.4 Phase diagram of the bag model, obtained setting $N_f = 2$ and using $m_u = m_d = 0$, $\mu_u = \mu_d = \mu$, and $B = 57.5 \, \text{MeV/fm}^3$, corresponding to $B^{1/4} = 145 \, \text{MeV}$. The hadronic phase is described as a pion gas with vanishing chemical potential

4.2.2 The Equation of State of Quark Matter

The thermodynamic functions describing a many-particle system at temperature $T = 1/\beta$ can be obtained from the grand partition function, defined as

$$Z = \mathrm{Tr}\exp\left[-\beta\left(H - \sum_i \mu_i N_i\right)\right],\tag{4.26}$$

where H is the hamiltonian operator and μ_i and N_i are the chemical potential and the number operator associated with particles of species i, respectively.

For example, pressure and energy density—denoted, respectively, P and ϵ—are obtained from the grand canonical potential Ω, which is in turn related to Z through

$$\Omega = -\frac{1}{\beta}\ln Z,\tag{4.27}$$

using

$$P = -\frac{\Omega}{V},\tag{4.28}$$

and

$$\epsilon = \Omega - \sum_i \left(\frac{\partial\Omega}{\partial\mu_i}\right)_{T,V},\tag{4.29}$$

V being the normalisation volume.

Owing to the properties of QCD, in the case of quark matter Ω consists of two contributions. One of them, that will be denoted Ω_{pert}, is tractable in perturbation theory, while the other one takes into account non perturbative effects originating by the properties of the QCD vacuum.

The relation between pressure and grand canonical potential, Eq. (4.28), combined with the interpretation of the bag constant discussed in Sect. 4.2.1, suggests that within the MIT bag model the difference between Ω and Ω_{pert} can be identified with the bag constant, which implies that

$$\Omega = \Omega_{\mathrm{pert}} + VB,\tag{4.30}$$

The perturbative contribution can be expanded according to

$$\Omega_{\mathrm{pert}} = V\sum_f\sum_n \Omega_f^{(n)}\tag{4.31}$$

where the index f specifies the quark flavour, while $\Omega_f^{(n)}$ is the n-th order term of the perturbative expansion in powers of the strong coupling constant, α_s.

The EOS of quark matter can be obtained from the relations linking pressure and energy density to Ω

$$P = -\frac{\Omega}{V} = -B - \sum_f \sum_n \Omega_f^{(n)} . \tag{4.32}$$

and

$$\epsilon = \frac{\Omega}{V} - \sum_f \left(\frac{\partial \Omega}{\partial \mu_f} \right) = -P + \sum_f \mu_f n_f , \tag{4.33}$$

n_f and μ_f being the density and chemical potential of the quarks of flavor f.

The lowest order contribution at $T = 0$, to be compared to Eq. (1.30), is given by

$$\Omega_f^{(0)} = -\frac{1}{\pi^2} \left[\frac{1}{4} \mu_f \sqrt{\mu_f^2 - m_f^2} \left(\mu_f^2 - \frac{5}{2} m_f^2 \right) \right.$$
$$\left. + \frac{3}{8} m_f^4 \log \frac{\mu_f + \sqrt{\mu_f^2 - m_f^2}}{m_f} \right] , \tag{4.34}$$

where m_f is the quark mass.

Substituting Eq. (4.34) into Eqs. (4.32) and (4.33) and taking the limit of massless quarks one arrives at the EOS

$$P = \frac{\epsilon - 4B}{3} . \tag{4.35}$$

The contribution of first order in α_s, arising from one-gluon exchange processes, turns out to be

$$\Omega_f^{(1)} = \frac{2\alpha_s}{\pi^3} \left[\frac{3}{4} \left(\mu_f \sqrt{\mu_f^2 - m_f^2} - m_f^2 \log \frac{\mu_f + \sqrt{\mu_f^2 - m_f^2}}{m_f} \right)^2 - \frac{1}{2} (\mu_f^2 - m_f^2)^2 \right]. \tag{4.36}$$

The chemical potentials appearing in Eqs. (4.34) and (4.36) can be written in the form

$$\mu_f = e_{F_f} + \delta \mu_f = \sqrt{m_f^2 + p_{F_f}^2} + \delta \mu_f , \tag{4.37}$$

where the first term is the Fermi energy of a gas of noninteracting quarks of mass m_f at density $n_f = p_{F_f}^3/\pi^2$, whereas the second term is the perturbative correction of order α_s, whose explicit expression is [85]

$$\delta\mu_f = \frac{2\alpha_s}{3\pi^2}\left[p_{F_f} - 3\frac{m_{F_f}^2}{e_{F_f}}\log\left(\frac{e_{F_f} + p_{F_f}}{m_f}\right)\right].$$ (4.38)

Including both $\Omega_f^{(0)}$ and $\Omega_f^{(1)}$ in Eqs. (4.32) and (4.33) and taking again the limit of vanishing quark masses one finds

$$P = \frac{1}{4\pi^2}\left(1 - \frac{2\alpha_s}{\pi}\right)\sum_f \mu_f^4 - B,$$ (4.39)

$$\epsilon = \frac{3}{4\pi^2}\left(1 - \frac{2\alpha_s}{\pi}\right)\sum_f \mu_f^4 + B.$$ (4.40)

Comparison to Eq. (4.35) shows that, at first order of perturbation theory, the EOS of a system of massless quarks turns out to be unaffected by one-gluon exchange interactions.

For any baryon density, quark densities are dictated by the requirements of baryon number conservation, charge neutrality and weak equilibrium. In the case of two flavours, in which only the light up and down quarks are present, they imply the relations

$$n_B = \frac{1}{3}(n_u + n_d),$$ (4.41)

$$\frac{2}{3}n_u - \frac{1}{3}n_d - n_e = 0$$ (4.42)

$$\mu_d = \mu_u + \mu_e,$$ (4.43)

where n_e and μ_e denote the density and chemical potential of the electrons produced by the reaction

$$d \rightarrow u + e^- + \bar{\nu}_e.$$ (4.44)

Note that we have not taken into account the possible appearance of muons, as in the density region relevant to neutron stars μ_e never exceeds the muon mass.

As baryon density increases, the d-quark chemical potential reaches the value $\mu_d = m_s$, m_s being the mass of the strange quark. The energy of quark matter can then be lowered turning d-quarks into s-quarks through

$$d + u \rightarrow u + s,$$ (4.45)

and, in the presence of three flavors, Eqs. (4.41)–(4.43) become

$$n_B = \frac{1}{3}(n_u + n_d + n_s),\tag{4.46}$$

$$\frac{2}{3}n_u - \frac{1}{3}n_d - \frac{1}{3}n_s - n_e = 0,\tag{4.47}$$

$$\mu_d = \mu_s = \mu_u + \mu_e.\tag{4.48}$$

Unfortunately, the parameters entering the bag model EOS are only loosely constrained by phenomenology, and their choice involves a large uncertainty.

Because quarks are confined and not observable as individual particles, their masses are not directly measurable and must be inferred from hadron properties. The 2020 edition of the Review of Particle Physics reports the values $m_u = 2.16^{+0.49}_{-0.26}$ MeV $m_d = 4.67^{+0.48}_{-0.17}$ MeV and $m_s = 93.4^{+8.6}_{-3.4}$ MeV [86]. At typical neutron stars densities heavier quarks do not play a role.

The strong coupling constant α_s can be obtained from the renormalisation group equation

$$\alpha_s = \frac{12\pi}{(33 - 2N_f)\ln(\bar{\mu}^2/\Lambda^2)},\tag{4.49}$$

where $N_f = 3$ is the number of active flavours, Λ is the QCD scale parameter and $\bar{\mu}$ is an energy scale typical of the relevant density region, such as, e.g., the average quark chemical potential. Using $\Lambda \sim 100 \div 200$ MeV and setting $\bar{\mu} = \mu_d \sim \mu_u$ at a reference baryon density $n_B \sim 4n_0$ one gets $\alpha_s \sim 0.4 \div 0.6$.

The values of the bag constant resulting from fits of the hadron spectrum range between $\sim$57 MeV fm^{-3}, with $\Lambda = 220$ MeV, and $\sim$350 MeV fm^{-3}, with $\Lambda = 172$ MeV. However, the requirement that the deconfinement transition does not occur at density $\sim n_0$ constrains B to be larger than $\sim$120–150 MeV fm^{-3}, and the results of lattice calculations suggest a value of $\sim$210 MeV fm^{-3}.

Figure 4.5 shows the energy density of charge-neutral quark matter in weak equilibrium as a function of baryon density, for different values of B and α_s. Comparison between the dotted line, obtained setting $\alpha_s = 0$, and those corresponding to $\alpha_s \neq 0$ shows that perturbative gluon exchange produces a sizable change of slope, that cannot be obtained by adjusting the value of the bag constant.

The composition of charge neutral quark matter in weak equilibrium predicted from the MIT bag model is shown in Fig. 4.6. Note that at large densities quarks of the three different flavors are present in equal number, and leptons are no longer necessary to fulfil the requirement of charge neutrality.

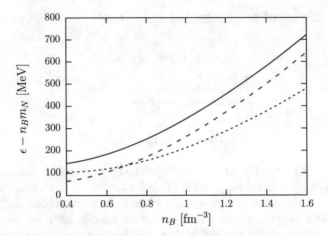

Fig. 4.5 Energy density of charge-neutral quark matter in weak equilibrium as a function of baryon number density. The solid and dashed lines have been obtained setting $\alpha_s = 0.5$, and $B = 200$ and $120\,\text{MeV}\,\text{fm}^{-3}$, respectively, while the dotted line corresponds to $\alpha_s = 0$ and $B = 200\,\text{MeV}\,\text{fm}^{-3}$. The quark masses are $m_u = m_d = 0$, and $m_s = 150\,\text{MeV}$

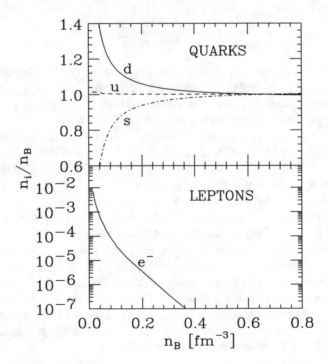

Fig. 4.6 Composition of charge neutral matter of u, d and s quarks and electrons in weak equilibrium, obtained from the MIT bag model setting $m_u = m_d = 0$, $m_s = 150\,\text{MeV}$, $B = 200\,\text{MeV/fm}^3$ and $\alpha_s = 0.5$

4.3 The Nuclear Matter-Quark Matter Phase Transition

At large baryon number density, the energy density of nucleon matter predicted by theoretical calculations, carried out within either NMBT or RMF theory, grows according to $\epsilon_{NM} \propto n^2$. On the other hand, the energy density of quark matter grows according to $\epsilon_{NM} \propto n^{4/3}$. Hence, at large enough density quark matter is expected to become the ground state of matter. If the density $\bar{n}_B$ corresponding to $\epsilon_{NM}(\bar{n}_B) = \epsilon_{QM}(\bar{n}_B)$ is reached in the inner core, the neutron star contains deconfined quark matter.

Early studies of the appearance of quark matter in neutron stars were based on the familiar Maxwell double tangent construction, which amounts to assuming that the transition occurs at constant pressure. Within this picture, charge-neutral nuclear matter at energy density ϵ_{NM} coexists with charge-neutral quark matter at energy density ϵ_{QM}, the two phases being separated by a sharp interface.

In the early 1990s Glendenning first pointed out that the requirement that the two phases be individually charge-neutral is, in fact, too restrictive. In a more general scenario, charged nuclear and quark matter may share a common lepton background, giving rise to a *mixed phase* extending over a sizeable fraction of the star [87].

4.3.1 Coexisting Phases vs Mixed Phase

Equilibrium between charged phases of nuclear matter and quark matter at $T = 0$ requires the fulfilment of Gibbs conditions

$$P_{NM}(\mu_{NM}^B, \mu_{NM}^Q) = P_{QM}(\mu_{QM}^B, \mu_{QM}^Q), \tag{4.50}$$

$$\mu_{NM}^B = \mu_{QM}^B \quad, \quad \mu_{NM}^Q = \mu_{QM}^Q, \tag{4.51}$$

where P denotes the pressure, while μ^B and μ^Q are the chemical potentials associated with the two conserved quantities, namely baryon number B and electric charge Q.

The above equations imply that, for any pressure $\bar{P}$, the projection of the surfaces $P_{NM}(\mu^B, \mu^Q)$ and $P_{QM}(\mu^B, \mu^Q)$ onto the $P = \bar{P}$ plane defines two curves, whose intersection corresponds to the equilibrium values of the chemical potentials. Since the chemical potentials determine the charge densities of the two phases, the volume fraction occupied by quark matter, χ, can then be obtained exploiting the requirement of *global* neutrality

$$\chi Q_{QM} + (1 - \chi)Q_{NM} + \sum_{\ell} Q_{\ell} = 0, \tag{4.52}$$

where Q_{QM}, Q_{NM} and Q_ℓ denote the electric charge carried by nuclear matter, quark matter and leptons, respectively. From Eq. (4.52) it follows that

$$\chi = \frac{Q_{NM} + \sum_\ell Q_\ell}{Q_{NM} - Q_{QM}}, \tag{4.53}$$

with $0 \le \chi \le 1$. Finally, the total energy density ϵ can be calculated using

$$\epsilon = \chi \epsilon_{QM} + (1 - \chi) \epsilon_{NM}, \tag{4.54}$$

and the EOS of state of the mixed phase can be cast in the standard form $P = P(\epsilon)$.

Requiring that the two phases be individually neutral, as in the pioneering work of Baym and Chin [88], reduces the number of chemical potentials to one, thus leading to the equilibrium conditions

$$P_{NM}(\mu_{NM}^B) = P_{QM}(\mu_{QM}^B), \tag{4.55}$$

$$\mu_{NM}^B = \mu_{QM}^B. \tag{4.56}$$

Within this scenario, charge-neutral nuclear matter at baryon number density n_B^{NM} coexists with charge-neutral quark matter at density n_B^{QM}, n_B^{NM} and n_B^{QM} being determined by the requirements

$$\mu_B = \left(\frac{\partial \epsilon_{NM}}{\partial n_B}\right)_{n_B = n_B^{NM}} = \left(\frac{\partial \epsilon_{QM}}{\partial n_B}\right)_{n_B = n_B^{QM}}. \tag{4.57}$$

At $n_B^{NM} < n_B < n_B^{QM}$ pressure and chemical potential remain constant, the energy density is given by

$$\epsilon = \mu_B n_B - P, \tag{4.58}$$

and the volume fraction occupied by quark matter grows linearly with density according to

$$\chi = \frac{\mu_B n_B - P - \epsilon_{NM}(n_{NM}^B)}{\epsilon_{QM}(n_B^{QM}) - \epsilon_{NM}(n_B^{NM})}. \tag{4.59}$$

Note that the above equation is obviously consistent with the requirement $0 \le \chi \le 1$, with $\chi(n_B^{NM}) = 0$ and $\chi(n_B^{QM}) = 1$.

The intersection between the surfaces describing the pressure of nuclear and quark matter can be determined numerically choosing as independent variables, instead of μ^B and μ^Q, the proton and neutron chemical potentials μ_p and μ_n. In nuclear matter they are simply related to the lepton chemical potential through the

β-stability condition $\mu_n - \mu_p = \mu_\ell$. In quark matter the chemical potentials of up and down quarks can be obtained from μ_p and μ_n, inverting the relations

$$\mu_p = 2\mu_u + \mu_d, \tag{4.60}$$

$$\mu_n = 2\mu_d + \mu_u, \tag{4.61}$$

while the strange quark and lepton chemical potentials are dictated by the conditions of weak equilibrium

$$\mu_s = \mu_d, \tag{4.62}$$

$$\mu_d - \mu_u = \mu_\ell. \tag{4.63}$$

Figure 4.7 illustrates the construction employed to determine the values of μ_p and μ_n corresponding to equilibrium between nuclear matter, described by NMBT, and quark matter, described by by the MIT bag model EOS with $\alpha_s = 0.5$ and $B = 200\,\text{MeV/fm}^3$. The region $P_{min} < P < P_{max}$ in which the isobars of nuclear and quark matter intersect defines the range of densities $n_{min} < n_B < n_{max}$ in which the mixed phase is energetically favored. At $n_B < n_{min}$ and $n_B > n_{max}$ the ground state consists of pure nuclear and quark matter, respectively.

The phase transition between nuclear and quark matter, obtained with $B = 200\,\text{MeV/fm}^3$ and $\alpha_s = 0.5$, is illustrated in Fig. 4.8. Dashed and dotdash lines show the dependence on n_B on the energy density of charge neutral nuclear and quark matter in weak equilibrium, respectively, while the solid line corresponds to the mixed phase. The latter turns out to be the ground state of neutron star matter at densities $0.7 \lesssim n_B \lesssim 1.7\,\text{fm}^{-3}$.

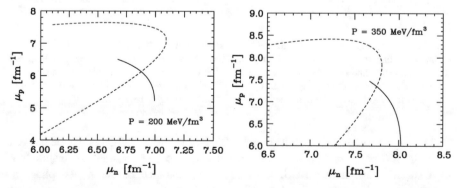

Fig. 4.7 Isobars $P(\mu_n, \mu_p) - 200\,\text{MeV/fm}^3$ (left panel) and $350\,\text{MeV/fm}^3$ (right panel) obtained using NMBT (solid lines) and the MIT bag model of quark matter, with $\alpha_s = 0.5$ and $B = 200\,\text{MeV/fm}^3$ (dashed lines). The intersections determine the values of the chemical potentials corresponding to equilibrium of the two phases according to Gibbs rules. Reprinted from Ref. [89] with permissions, © ESO 2005. All rights reserved

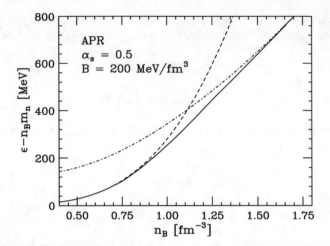

Fig. 4.8 Dashed and dotdash lines show the energy density of charge neutral nuclear and quark matter in weak equilibrium, respectively. The bag model parameters have been set to $B = 200 \, \mathrm{MeV/fm^3}$ and $\alpha_s = 0.5$. The solid line corresponds to the mixed phase, obtained from Gibbs equilibrium conditions. Reprinted from Ref. [89] with permissions, © ESO 2005. All rights reserved

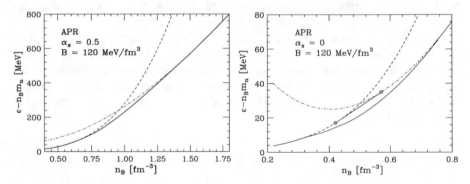

Fig. 4.9 As in Fig. 4.8, but with the EOS of quark matter obtained using different values of the MIT bag model parameters. Left panel: $B = 120 \, \mathrm{MeV/fm^3}$ and $\alpha_s = 0.5$; right panel $B = 120 \, \mathrm{MeV/fm^3}$ and $\alpha_s = 0$. The straight line in the right panel is the double tangent obtained from the Maxwell construction, while the open circles show the extrema of the coexistence region. Reprinted from Ref. [89] with permissions, © ESO 2005. All rights reserved

The dependence of the results on the MIT bag model parameters can be gauged from the left panel of Fig. 4.9. A lower value of the bag constant, corresponding to a softer quark matter EOS, leads to the appearance of the mixed phase at lower density. Keeping $\alpha_s = 0.5$ and setting $B = 120 \, \mathrm{MeV/fm^3}$ the mixed phase turns out to be energetically favoured in the range $0.6 \lesssim n_B \lesssim 1.4 \, \mathrm{fm^{-3}}$. An even larger effect, illustrated by the right panel of Fig. 4.9 is obtained with $B = 120 \, \mathrm{MeV/fm^3}$ and $\alpha_s = 0$, which corresponds to neglecting perturbative gluon exchange altogether. For this case the figure also shows the results obtained from the

Maxwell construction, leading to the coexistence of charge-neutral nuclear matter at $n_B = 0.42$ fm^{-3} and charge-neutral quark matter at $n_B = 0.57$ fm^{-3}. This coexistence region is to be compared to the region of stability of the mixed phase, corresponding to $0.22 \lesssim n_B \lesssim 0.75$ fm^{-3}.

4.3.2 Stability of the Mixed Phase

The discussion of the previous section suggests that, irrespective of the details of the EOS, the transition from nuclear to quark matter is likely to proceed through the formation of a mixed phase. However, two issues relevant to both the appearance and stability of the mixed phase, not taken yet into account, need to be further analysed.

Consider a mixed phase consisting of droplets of quark matter immersed in β-stable nuclear matter, with global charge neutrality being guaranteed by a lepton background. This picture is obviously based on the assumption that the appearance of the charged droplets do not significantly affect the space distribution of the leptons, i.e. that the Debye screening length λ_D is large compared to both the typical size of the droplets and their separation distance. If this condition is not satisfied the lepton background is distorted in such a way as to screen electrostatic interactions.

Estimates of λ_D reported in the literature suggest that screening effects can be disregarded if the structures appearing in the mixed phase of quark and nuclear matters have typical size and separation distance of ~10 fm; see Ref. [90].

The second issue deserving consideration is the stability of the mixed phase, i.e. the question of whether or not its energy is lower than the energy of the coexisting phases of nuclear and quark matter.

The formation of a spherical droplet of quark matter requires the energy

$$\mathcal{E}_D \doteq \mathcal{E}_C + \mathcal{E}_S . \tag{4.64}$$

where the surface contribution $\mathcal{E}_S$ is parametrized in terms of the surface tension σ according to

$$\mathcal{E}_S = \sigma \, 4\pi R^2 , \tag{4.65}$$

R being the droplet radius. The electrostatic energy $\mathcal{E}_C$ can be cast in the form

$$\mathcal{E}_C = \frac{3}{5} \frac{Q^2}{R} \left(1 - \frac{3}{2} u^{1/3} + \frac{1}{2} u \right) , \tag{4.66}$$

with $u = (R/R_c)^3$, R_c being the radius of the Wigner-Seitz cell. Note that the first term on the right hand side of the above equation is the self energy of a droplet of radius R and charge Q obtained from Gauss law. The electric charge Q is given by

$$Q = \frac{4\pi R^3}{3} \left(\rho_{QM} - \rho_{NM} \right) , \tag{4.67}$$

ρ_{QM} and ρ_{NM} being the charge densities of quark matter and nuclear matter, respectively. Minimisation of the energy density $\epsilon = 3\mathcal{E}_D/4\pi R_c^3$ with respect to the droplet radius yields

$$\mathcal{E}_S = 2\mathcal{E}_C, \tag{4.68}$$

and

$$R = \left[\frac{4\pi (\rho_{QM} - \rho_{NM})^2}{3\sigma} f_3(u) \right]^{-1/3}, \tag{4.69}$$

where

$$f_3(u) = \frac{1}{5}\left(2 - 3u^{1/3} + u\right). \tag{4.70}$$

As the density increases, the droplets begin to merge and give rise to structures of variable dimensionality, changing first from spheres into rods and eventually into slabs.[2] At larger densities the volume fraction occupied by quark matter exceeds 1/2, and the role of the two phases is reversed. Nuclear matter, initially arranged in slabs, turns into rods and spheres that finally dissolve in uniform charge-neutral quark matter.

The energy density needed for the formation of the structures appearing in the mixed phase has been obtained by Ravenhall et al. in the context of a study of matter in the neutron star inner crust [24]. It can be written in the concise form

$$\epsilon_C + \epsilon_S = 6\pi u \left[\frac{\alpha}{4\pi}\sigma^2 d^2 \left(n_B^{NM} - n_B^{QM}\right)^2 f_d(u) \right], \tag{4.71}$$

where α is now the fine structure constant, u is the volume fraction occupied by the less abundant phase—that is, $u = \chi$ for $\chi < 1/2$, $u = 1 - \chi$ for $\chi \geq 1/2$—and

$$f_d(u) = \frac{1}{d+2}\left[\frac{1}{d-2}\left(2 - du^{1-2/d}\right) + u \right]. \tag{4.72}$$

For $d = 1, 2$ and 3 Eqs. (4.71) and (4.72) provide the energy density for the case of slabs, rods and spheres, respectively.

For $\sigma = 0$ both surface and Coulomb energies vanish, and the energy density of the mixed phase is given by Eq. (4.54), while for $\sigma \neq 0$

$$\epsilon(\sigma) = \epsilon(\sigma = 0) + \epsilon_C + \epsilon_S. \tag{4.73}$$

---

[2] These structures are similar to those occurring in the mixed phase of the neutron star inner crust, referred to as *pasta phase*; see Chap. 2.

The mixed phase is energetically favorable if $\epsilon(\sigma)$ is less than the energy density obtained from the Maxwell construction, given by Eq. (4.58).

The value of the surface tension at the interface between nuclear and quark matter is not known. It has been estimated using the MIT bag model and neglecting gluon exchange. Assuming that the strange quark has mass of $\sim$150 MeV, Berger and Jaffe obtained $\sigma \sim$ 10 MeV/fm^2 [91]. The stability of the mixed phase has been investigated by computing $\Delta_\epsilon = \epsilon(\sigma) - \epsilon(0)$ for different values of σ, ranging from 2 MeV/fm^2 to 10 MeV/fm^2 [89].

For any given value of the baryon number density n_B, the energy density of Eqs. (4.71) and (4.72) has been calculated using the nuclear and quark matter densities determined according to the procedure described in the previous section and carrying out a numerical minimization with respect to the value of the dimensionality parameter d. As n_B increases, the resulting values of d change initially from $\sim$3 to $\sim$2 and $\sim$1 and then again to $\sim$2 and finally to $\sim$3. For example, in the case illustrated by Fig. 4.10, and corresponding to $\sigma = $ 10 MeV/fm^2, spherical droplets of quark matter ($d \sim 3$) appear at $n_B \sim 0.75$ fm^{-3} and turn into rods ($d \sim 2$) and slabs ($d \sim 1$) at $n_B \sim 0.95$ and $\sim$1.2 fm^{-3}, respectively. For larger densities, quark matter becomes the dominant phase, implying $\chi > 1/2$: at $n_B \sim 1.5$ and $\sim$1.7 fm^{-3} the mixed phase features rods ($d \sim 2$) and droplets ($d \sim 3$) of nuclear matter that eventually dissolve in the quark matter background.

These results are summarised in Figs. 4.10 and 4.11, corresponding to different choices of the MIT bag model parameters. The solid lines show the n_B dependence

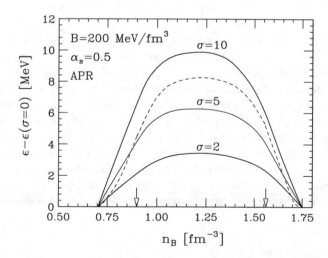

Fig. 4.10 The solid lines correspond to the difference $\Delta_\epsilon = \epsilon(\sigma) - \epsilon(0)$—see Eq. (4.73)—evaluated for $\sigma = $ 10, 5 and 2 MeV/fm^3. The dashed line shows the difference $\tilde{\Delta}$ between the energy density resulting from the Maxwell construction and $\epsilon(0)$. The arrows mark the limits of the coexistence region. Nuclear and quark matter are described by the NMBT and the MIT bag model EOS, with $\alpha_s = 0.5$ and $B = $ 200 MeV/fm^3, respectively. Reprinted from Ref. [89] with permissions, © ESO 2005. All rights reserved

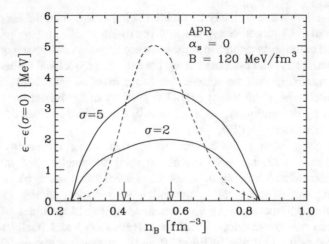

Fig. 4.11 Same as in Fig. 4.10, but with $\alpha_s = 0$ and $B = 120\,\mathrm{MeV/fm^3}$. The solid lines show to the difference $\Delta_\epsilon = \epsilon(\sigma) - \epsilon(0)$ (see Eq. (4.73)), evaluated for $\sigma = 5$ and $2\,\mathrm{MeV/fm^3}$, respectively. Reprinted from Ref. [89] with permissions, © ESO 2005. All rights reserved

of the calculated Δ_ϵ for different values of the surface tension σ. The dashed line represents the difference $\widetilde{\Delta}_\epsilon = \epsilon_M - \epsilon(0)$, where ϵ_M is the energy density obtained from the Maxwell construction. For any given value of the surface tension, the mixed phase is favorable if the corresponding solid line lies below the dashed line.

The results of Fig. 4.10, corresponding to $B = 200\,\mathrm{MeV/fm^3}$ and $\alpha_s = 0.5$, show that the mixed phase, while being always the lowest energy phase for $\sigma = 2\,\mathrm{MeV/fm^2}$, becomes energetically unfavorable at some densities for $\sigma \gtrsim 5\,\mathrm{MeV/fm^2}$. For $\sigma = 10\,\mathrm{MeV/fm^2}$ coexistence of charge neutral phases of nuclear and quark matter turns out to be favorable over the whole density range.

The dependence on the MIT bag model parameters can be gauged from Fig. 4.11, showing results obtained setting $B = 120\,\mathrm{MeV/fm^3}$ and $\alpha_s = 0$. It appears that for σ in the range 2–5 MeV/fm² the mixed phase is energetically favourable over a density region larger than the coexistence region predicted by the Maxwell construction.

Returning to the issue of Debye screening, it should be pointed out that in the region of $\chi \ll 1$, corresponding to formation of droplets of quark matter, the droplets radius obtained setting $B = 200\,\mathrm{MeV/fm^3}$ and $\alpha_s = 0.5$ turns out to be $\sim$2–3 fm. This value suggest that the conditions derived by Heiselberg et al. [90] are indeed fulfilled.

4.3.3 Strange Stars

Following the pioneering study of the transition from neutron matter to quark matter carried out by Baym and Chin using the MIT bag model, a great deal of effort has been devoted to the development of more advanced treatments, based on the model of hadron dynamics that Nambu and Jona-Lasinio (NJL) originally proposed in the 1960s [92]. The main assumption underlying their work was that at low energy and momentum the gluon degrees of freedom are frozen, and quark dynamics can be described by a local effective interaction. While not including a mechanism leading to quark confinement, the NJL model explains the spontaneous breakdown of chiral symmetry, and provides a quantitative account of the spectrum of pseudoscalar mesons.

With the advent of QCD as the fundamental theory of strong interactions, the NJL model was largely set aside until the 1980s, when it was revived as an effective theory of strongly interacting matter in the non perturbative regime. A most appealing feature of this approach is the capability to describe a variety of colour superconducting phases, which are believed to be the ground state of degenerate quark matter at high density. In analogy with the case of superfluid neutron matter discussed in Sect. 2.2.2, the stability of the superconducting phase is due to the formation of pairs of quarks bound by an attractive colour interaction. The application of the NJL model to study the properties of dense quark matter is extensively analysed in Ref. [93].

It has been also suggested that, in addition to occupying the innermost region of *hybrid stars*, quark matter may form a new family of compact stars consisting almost entirely of deconfined quarks. Following a pioneering work of Bodmer [94], in 1984 Witten argued that matter comprising up, down and strange quarks with strangeness per baryon of order unity, or *strange matter*, may, in fact, be *absolutely stable* [95]. According to this hypothesis, *quark stars*, unlike ordinary neutron stars, are *self bound*, that is, bound even in the absence of gravitational attraction. Owing to this character, these astrophysical objects, also referred to as *strange stars*, are predicted to exhibit distinct, and potentially observable properties, such as a small radius and fast cooling rate.

Appendix: Partition Function of Fermion Systems

The study of the different phases of QCD at non-zero temperature and chemical potentials involves the thermodynamic functions, that can be obtained from the grand canonical partition function Z. For a system at temperature $T = \beta^{-1}$

$$Z = \mathrm{Tr}\, \exp[-\beta(H - \sum_i \mu_i N_i)], \tag{4.74}$$

where H is the Hamiltonian describing the microscopic dynamics, the N_i are number operators associated with conserved quantities—such as, e.g., the difference between the numbers of electrons and positrons in the case of QED—and the μ_i are the corresponding chemical potentials.

From Eq. (4.74) it follows that

$$P = -T \frac{\partial \ln Z}{\partial V} , \tag{4.75}$$

$$N_i = T \frac{\partial \ln Z}{\partial \mu_i} , \tag{4.76}$$

$$S = \frac{\partial T \ln Z}{\partial T} , \tag{4.77}$$

$$E = -PV + TS + \sum_i \mu_i N_i = T \frac{\partial \ln Z}{\partial T} + T \sum_i \mu_i \frac{\partial \ln Z}{\partial \mu_i} , \tag{4.78}$$

where P, S and E denote the pressure, entropy and energy of the system, respectively.

Equations (4.75)–(4.78) clearly show that it is convenient to introduce the thermodynamic potential

$$\Omega = -T \ln Z = -PV , \tag{4.79}$$

the partial derivatives of which yield the thermodynamic functions.

In quantum field theory, the trace operation appearing in the definition of Z, that is, the sum of all diagonal elements of the density matrix, is performed exploiting the formalism of path integrals. Consider, as an example, the simple case of a scalar field theory, the Lagrangian density of which reads

$$\mathcal{L} = \frac{1}{2} \partial^\mu \partial_\mu \phi - \frac{1}{2} m^2 \phi^2 . \tag{4.80}$$

Because there are no conserved charges, the partition function reduces to

$$Z = N \int \mathcal{D}\phi \, e^S , \tag{4.81}$$

with the integration being extended to all periodic paths such that $\phi(\mathbf{x}, \tau) = \phi(\mathbf{x}, 0)$, and the imaginary time action is given by

$$S = \int_0^\beta d\tau \int d^3x \, \mathcal{L}(\phi, \partial\phi/\partial\tau)$$

$$= -\frac{1}{2} \int_0^\beta d\tau \int d^3x \left[\left(\frac{\partial\phi}{\partial\tau} \right)^2 + (\nabla)\phi)^2 + m^2\phi^2 \right] . \tag{4.82}$$

with $\tau = it$. Note that the normalisation factor appearing in the right-hand side of Eq. (4.81) is not relevant in this context, because the thermodynamic functions are not affected by multiplicative factors.

The calculation of the partition function of Dirac's theory, described in detail in Sect. 4.2.1, yields the result

$$Z = \int i\mathcal{D}\psi^\dagger \int \mathcal{D}\psi \, \exp \int_0^\beta d\tau \int d^3x \, \bar{\psi} \left(-\gamma^0 \frac{\partial}{\partial \tau} + i\boldsymbol{\gamma} \cdot \boldsymbol{\nabla} - m_i + \mu_i \gamma^0 \right) \psi \,,$$

(4.83)

where $i\psi^\dagger$ and ψ are independent dynamical variables, and the integration is extended to all antiperiodic paths, fulfilling the condition $\psi(\mathbf{x}, 0) = -\psi(\mathbf{x}, \beta)$, as required by the anticommutation rules obeyed by fermionic fields.

Neutrino Emission from Neutron Stars

5

Abstract

The temperature of a newly formed neutron star is believed to be as high as 10^{11} K. After a time $t \sim 50$ s, however, the neutrino mean free path becomes larger than the star radius, and the temperature begins to decrease due to the energy loss associated with neutrino emission processes. Depending on the properties of matter in the star interior, neutrino emission remains the dominant cooling mechanism for a time varying between few weeks and $\sim 10^6$ years. At the end of this epoch the temperature is reduced to $\sim 10^8$ K, and photon emission becomes dominant. This chapter provides a concise review of the neutrino emission processes taking place in the neutron star core, where matter consists mainly of neutrons, with a small fraction of protons and leptons determined by the requirements of β-equilibrium and charge neutrality.

5.1 Direct Urca Process

The most efficient neutrino and antineutrino production reactions, dubbed Urca processes, were first discussed by Gamow and Schoenberg in the 1940s [96].[1] These are the simplest weak interaction processes occurring in npe matter, that is, neutron β-decay and electron capture by protons

$$n \rightarrow p + e + \bar{\nu}_e , \qquad p + e \rightarrow n + \nu_e . \tag{5.1}$$

---

[1] According to Gamow's autobiography [97], the name was chosen "because the Urca Process results in a rapid disappearance of thermal energy from the interior of a star, similar to the rapid disappearance of money from the pockets of the gamblers on the Casino de Urca".

© The Author(s), under exclusive license to Springer Nature Switzerland AG 2023   107
O. Benhar, *Structure and Dynamics of Compact Stars*, Lecture Notes
in Physics 1019, https://doi.org/10.1007/978-3-031-35628-5_5

At equilibrium, the above reactions take place at the same rate, and the chemical potentials of the participating particles satisfy the condition[2]

$$\mu_n = \mu_p + \mu_e \,. \tag{5.2}$$

Because the rate of production of neutrinos and antineutrinos is the same, the total neutrino emissivity Q_D, defined as the neutrino and antineutrino energy emitted per unit time and volume, is simply twice the emissivity associated with neutron β-decay. Its expression reads

$$Q_D = 2 \int \frac{d\mathbf{p}_n}{(2\pi)^3} \, dW_{i \to f} \, E_\nu \, f_n \, (1 - f_p) \, (1 - f_e) \,. \tag{5.3}$$

In the above equation E_ν denotes the neutrino energy, f_j, with $j = n, p, e$, is the equilibrium Fermi-Dirac distribution of particles of species j, carrying momentum $\mathbf{p}_j$ and energy E_j, defined as

$$f_j = \frac{1}{1 + e^{(E_j - \mu_j)/T}} \,, \tag{5.4}$$

which μ_j being the chemical potential. The derivation of the differential probability of β-decay

$$dW_{i \to f} = 2\pi \delta(E_n - E_p - E_e - E_\nu)\delta(\mathbf{p}_n - \mathbf{p}_p - \mathbf{p}_e)$$
$$\times 2G^2 \left(1 + 3g_A{}^2\right) 4\pi E_\nu^2 \, dE_\nu \frac{d\mathbf{p}_p}{(2\pi)^3} \frac{d\mathbf{p}_e}{(2\pi)^3} \,, \tag{5.5}$$

is discussed in Appendix 1.

Note that, in principle, the calculation of Eq. (5.3) involves a ten-dimensional integration. However, it can be greatly simplified—and performed analytically—using the procedure known as *phase space decomposition*, which is expected to be accurate for strongly degenerate fermions. This scheme is based on the tenet that, because at the temperatures typical of neutron stars nucleons and leptons are both strongly degenerate, the main contribution to the emissivity originates from particles occupying quantum states in the vicinity of the Fermi surface. As a consequence, one can use the approximation $|\mathbf{p}_j| \approx p_{F_j}$, implying that the typical neutrino energy is $\sim T$. The neutrino momentum is also of order T, and can safely be neglected, with respect to the momenta of the degenerate fermions, in the argument of the momentum-conserving δ-function.

---

[2] Here, we assume that neutrinos and antineutrinos are non degenerate. As a consequence, they have vanishing density and chemical potential.

Using the above approximations, the differential transition probability can be cast in the form

$$dW_{i \to f} = 2\pi \, \delta(E_n - E_p - E_e - E_\nu) \, \delta(\mathbf{p}_n - \mathbf{p}_p - \mathbf{p}_e) \, |M_{fi}|^2 \tag{5.6}$$

$$\times \, 4\pi E_\nu^2 \, dE_\nu \, \frac{d\mathbf{p}_p}{(2\pi)^3} \, \frac{d\mathbf{p}_e}{(2\pi)^3} \, ,$$

with $|M_{fi}|^2 = 2G^2(1 + 3g_A^2)$. Here, $G = G_F \cos\theta_c$, G_F and θ_c being the Fermi constant and Cabibbo's angle, respectively, while g_A is the axial-vector coupling constant.

The above equation can be further rewritten applying phase-space decomposition. The resulting expression reads

$$dW_{i \to f} \, \frac{d\mathbf{p}_n}{(2\pi)^3} = \frac{1}{(2\pi)^8} \, \delta(E_n - E_p - E_e - E_\nu) \, \delta(\mathbf{p}_n - \mathbf{p}_p - \mathbf{p}_e) \tag{5.7}$$

$$\times \, |M_{fi}|^2 \, 4\pi E_\nu^2 \, dE_\nu \, \prod_{j=1}^{3} p_{Fj} \, m_j^* \, dE_j \, d\Omega_j \, ,$$

where $d\Omega_j$ is the differential solid angle specifying the direction of the momentum $\mathbf{p}_j$, $m_j^* = p_{Fj}/v_{Fj}$ is the effective mass of the particles of species j and $v_{Fj} = (\partial E_j / \partial p_j)_{p=p_{Fj}}$, denotes the corresponding Fermi velocity.

Substitution of Eq. (5.7) in the definition of the neutrino emissivity, Eq. (5.3), leads to the result

$$Q_D = \frac{2}{(2\pi)^8} \, T^6 \, A \, I \, |M_{fi}|^2 \, \prod_{j=1}^{3} p_{Fj} \, m_j^* \, , \tag{5.8}$$

where

$$A = 4\pi \int d\Omega_1 \, d\Omega_2 \, d\Omega_3 \, \delta(\mathbf{p}_n - \mathbf{p}_p - \mathbf{p}_e) \, , \tag{5.9}$$

and

$$I = \int_0^\infty dx_\nu \, x_\nu^3 \left[\prod_{j=1}^{3} \int_{-\infty}^\infty dx_j \, f_j \right] \delta(x_1 + x_2 + x_3 - x_\nu) \, . \tag{5.10}$$

The definition of A, Eq. (5.9), involves the integration over the angular variables, while the calculation of I, defined by Eq. (5.10), requires integrations over the dimensional variables $x_\nu = E_\nu/T$ and $x_j = (E_j - \mu_j)/T$. The details of the calculations of both A and I can be found in Appendix 2. The final results are

$$A = \frac{32\pi^3}{p_{Fn}\, p_{Fp}\, p_{Fe}}\, \Theta_{npe} \quad , \quad I = \frac{457\pi^6}{5040} , \tag{5.11}$$

where the function

$$\Theta_{npe} = \begin{cases} 1 & \text{if } p_{Fn} \le p_{Fp} + p_{Fe} \\ 0 & \text{otherwise} \end{cases} , \tag{5.12}$$

enforces the threshold condition that must be satisfied to fulfil the requirement of momentum conservation; see below. The resulting expression of the emissivity is

$$Q_D = \frac{457}{10,080}\, G_F^2 \cos^2 \theta_c\, (1 + 3g_A^2)\, m_n^* m_p^* m_e^*\, T^6\, \Theta_{npe}$$

$$\simeq 4.00 \times 10^{27} \left(\frac{n_e}{n_0}\right)^{1/3} \frac{m_n^* m_p^*}{m_n^2}\, T_9^6\, \Theta_{npe}\ \text{erg cm}^{-3}\ \text{s}^{-1} , \tag{5.13}$$

where $n_0 = 0.16\ fm^{-3}$ is the baryon number density of isospin symmetric nuclear matter at equilibrium—obtained from extrapolation of the central density of atomic nuclei and the semi empiric mass formula—while T_9 denotes the temperature measured in units of 10^9 K. Note that the temperature dependence of the emissivity is described by the power law $Q_D \propto T_9^6$.

The appearance of the sixth power of T can be easily explained using simple phase space considerations. Integration over the momenta of the three strongly degenerate fermions participating in the process contributes a term $\propto T^3$, as the integration region is limited to a thin shell of size $\sim T$ around the Fermi surface. On the other hand, the integration over the momentum of the non degenerate neutrino is not restricted, and its contribution is $\propto T^3$. Taking into account that the energy-conserving δ-function brings about a factor T^{-1} that cancels the factor T associated with the neutrino energy, one obtains the T^6 dependence of Eq. (5.13). This example clearly illustrates how degeneracy strongly affects the neutrino emissivity of neutron star matter.

5.1.1 Threshold of the Direct Urca Process

The main feature of the direct Urca process is the occurrence of a threshold, resulting in the appearance of the step function Θ_{npe} of Eq. (5.13). Because the neutrino emissivity associated with this process—when active—turns out to be dominant by many order of magnitude, it is very important to understand the implications of the threshold condition.

As pointed out above, the occurrence of the direct Urca process in β-stable npe matter requires that the Fermi momenta of the degenerate fermions, p_{Fn}, p_{Fp} e p_{Fe}, fulfill the triangular condition $p_{Fn} \leq p_{Fp} + p_{Fe}$. At densities around the equilibrium density of isospin symmetric matter, n_0, this requirement is *not* met. However, several models of the nuclear EOS predict that the Fermi momenta p_{Fp} and p_{Fe} can grow faster than p_{Fn} with increasing density, thus allowing the onset of the reaction.

The Urca process is active as soon as the proton fraction reaches a critical value. In the case of npe matter, charge neutrality implies $p_{Fp} = p_{Fe}$ and the triangular condition simplifies to $p_{Fp} \geq p_{Fn}/2$. It follows that the proton and neutron densities must be such that $n_p \geq n_n/8$. In terms of proton fraction this inequality corresponds to

$$x_p = \frac{n_p}{n_n + n_p} \geq \frac{1}{9} \, . \tag{5.14}$$

In the case of $npe\mu$ matter, if the electron chemical potential exceeds the muon rest mass m_μ, the direct Urca processes in which electrons are replaced by muons

$$n \rightarrow p + \mu + \bar{v}_\mu \, , \qquad p + \mu \rightarrow n + v_\mu \, , \tag{5.15}$$

are also energetically allowed, and must be taken into account. The emissivity associated with the above reactions can be written in the same form as Eq. (5.13), since the condition of β-equilibrium requires $\mu_\mu = \mu_e$, implying in turn $m_\mu^* = m_e^*$. The threshold function alone is modified to $\Theta_{np\mu}$, as the appearance of muons affects the critical value of the proton fraction, which grows until reaching the upper limit $x_p = 1/[1 + (1 + 2^{-1/3})^3] \approx 0.148$ when $\mu_\mu \gg m_\mu$.

As mentioned in the previous chapters, early models of neutron star matter were largely based on the non interacting Fermi gas description. Within this oversimplified picture, the proton fraction never exceeds the threshold of direct Urca processes. However, in the 1980s [98] and early 1990s [99] it was shown that the inclusion of the effects of strong interactions using dynamical models predicting large symmetry energies leads to a sizeable increase of the proton fraction. The resulting values of x_p turn out to be above the direct Urca threshold in a density region which is likely to be attained in the neutron star core.

Figure 5.1 shows the proton, electron and muon fractions in charge-neutral β-stable matter, obtained by Akmal et al. using a variational approach and the $A18 + \delta v + UIX'$ nuclear Hamiltonian discussed in Chap. 3 [63]. As previously pointed out, the kink at $n_B \lesssim 0.2$ fm^{-3} corresponds to the transition from the phase of normal nuclear matter to a phase featuring neutral pion condensation [62].

The solid and dashed lines of Fig. 5.2 represent the density dependence of the differences $p_{Fn} - p_{Fp} - p_{Fe}$ and $p_{Fn} - p_{Fp} - p_{F\mu}$ obtained using results reported in Ref. [63]. It is apparent that the triangular condition on the Fermi momenta is fulfilled at density above ~ 0.7 fm^{-3} and ~ 0.9 fm^{-3} for the electron and muon Urca process, respectively.

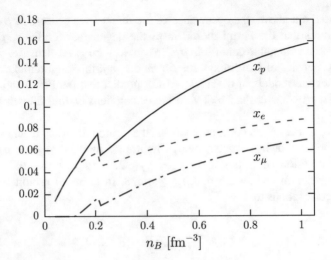

Fig. 5.1 Proton, electron and muon fractions of charge-neutral β-stable matter predicted by the model of Akmal et al. [63]

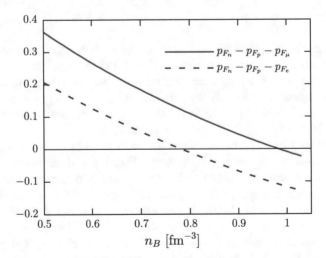

Fig. 5.2 Density dependence of the differences $p_{F_n} - p_{F_p} - p_{F_e}$ (dashed line) and $p_{F_n} - p_{F_p} - p_{F_\mu}$ (solid line), obtained from the model of Ref. [63]. The triangular conditions on the Fermi momenta are fulfilled at densities larger than the values at which the differences change sign

5.2 Modified Urca Processes

If the direct Urca process is not allowed, the most efficient mechanism of neutrino emission is the *modified* Urca process, in which momentum conservation is made possible by the presence of an additional nucleon, acting as a spectator.

The modified Urca reactions involve a nucleon-nucleon collision associated with β-decay or electron capture

$$n + n \rightarrow p + n + e + \bar{\nu}_e \ , \qquad p + n + e \rightarrow n + n + \nu_e \ , \tag{5.16}$$

$$n + p \rightarrow p + p + e + \bar{\nu}_e \ , \qquad p + p + e \rightarrow n + p + \nu_e \ . \tag{5.17}$$

These include the processes (5.16) and (5.17), referred to as neutron and proton branches, respectively. The addition of the spectator nucleon turns out to slow down the reaction rate dramatically.

In the following, modified Urca processes will be denoted by the label MN, where M stands for *modified*, whereas $N = n$ or p refers to the neutron or proton branch. Both branches include the direct and inverse reactions, the rates of which at equilibrium are the same. Hence, only one rate needs to be computed, and multiplied by a factor 2, to get the total emission rate.

Neutrino emissivity can be written in the form

$$Q^{MN} = 2 \int \left[\prod_{j=1}^{4} \frac{d\mathbf{p}_j}{(2\pi)^3} \right] \frac{d\mathbf{p}_e}{(2\pi)^3} \frac{d\mathbf{p}_\nu}{(2\pi)^3} E_\nu \, (2\pi)^4 \, \delta(E_f - E_i)$$

$$\times \ \delta(\mathbf{p}_f - \mathbf{p}_i) \, f_1 \, f_2 \, (1 - f_3) \, (1 - f_4) \, (1 - f_e) \frac{1}{2} |M_{fi}|^2 \ , \tag{5.18}$$

where the indices i and f refer to the initial and final states, and $|M_{fi}|^2$ is the squared transition amplitude of the process. The additional factor $1/2$ is a symmetry factor, the inclusion of which is needed to avoid double counting in collisions involving identical particles.

One can follow the same procedure employed in the discussion of the direct Urca process—see Appendix 2—and rewrite the emissivity associated with the modified reaction in the form

$$Q^{MN} = \frac{1}{(2\pi)^{14}} \, T^8 \, A \, I \, \langle |M_{fi}|^2 \rangle \prod_{j=1}^{5} p_{Fj} \, m_j^* \ , \tag{5.19}$$

with

$$A = 4\pi \left[\prod_{j=1}^{5} \int d\Omega_j \right] \delta(\mathbf{P}_f - \mathbf{P}_i) , \tag{5.20}$$

$$\langle |M_{fi}|^2 \rangle = \frac{4\pi}{A} \left[\prod_{j=1}^{5} \int d\Omega_j \right] \delta(\mathbf{P}_f - \mathbf{P}_i) |M_{fi}|^2 , \tag{5.21}$$

and

$$I = \int_0^\infty dx_\nu \, x_\nu^3 \left[\prod_{j=1}^{5} \int_{-\infty}^\infty dx_j \, f_j \right] \delta \left(\sum_{j=1}^{5} x_j - x_\nu \right) . \tag{5.22}$$

Note that, as the magnitudes of the momenta of the degenerate fermions, $\mathbf{p}_j$, are set to be equal to the corresponding Fermi momenta, the quantities A and $\langle |M_{fi}|^2 \rangle$ only involve integrations over the angles specifying the directions of the particle momenta.

In the evaluation of A, the integration over the neutrino momentum can be readily carried out, as $\mathbf{p}_\nu$ is neglected in the δ-function. Because, in general, $|M_{fi}|^2$ depends on the momenta, we have introduced the average $\langle |M_{fi}|^2 \rangle$.

A calculation of the integrals of Eq. (5.22) following the procedure described in Appendix 2 yields the result

$$I = \frac{11{,}513 \, \pi^8}{120{,}960} . \tag{5.23}$$

From this point on, the neutron and proton branches require separate discussions.

5.2.1   Neutron Branch

Let us consider the first reaction of (5.16), represented by the diagram of Fig. 5.3. The neutrons in the initial state are labelled 1 and 2, while 3 and 4 denote the neutron and proton in the final state, respectively. The calculation of A yields the result

$$A_n = \frac{2\pi (4\pi)^4}{p_{Fn}^3} . \tag{5.24}$$

The evaluation of the matrix element M_{fi} involves non trivial difficulties, because it includes strong interactions. The nucleon-nucleon potential, discussed in Chap. 3, features a long range component arising from OPE processes, taken into

Fig. 5.3 Diagrammatic representation of a modified Urca process belonging to the neutron branch (5.16). The dashed and wavy lines represent the nucleon-nucleon and weak interaction, respectively

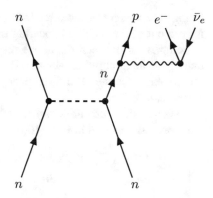

account within Yukawa's theory, and a short range, strongly repulsive, component, that is generally described within phenomenological approaches.

The long-range contribution has been estimated in the classic work of Friman and Maxwell [100], in which nucleons are treated as non relativistic particles and all lepton momenta are assumed to be negligible. The result, obtained averaging over the direction of the neutrino momentum, is written in the form

$$|M_{fi}^{MN}|^2 = \frac{16\,G^2}{E_e^2}\left(\frac{f}{m_\pi}\right)^4 g_A^2\,F_A\,,\qquad(5.25)$$

where m_π is the pion mass, and $f = g\,m_\pi/4m \approx 1$, with g being the pion-nucleon coupling constant appearing in the Lagrangian density of Yukawa's model, discussed in Chap. 3. The only non vanishing contribution to the right-hand side of Eq. (5.25), originating from the processes involving the axial-vector current, reads

$$F_A = \frac{4Q_1^4}{(Q_1^2+m_\pi^2)^2} + \frac{4Q_2^4}{(Q_2^2+m_\pi^2)^2} + \frac{(\mathbf{Q}_1\cdot\mathbf{Q}_2)^2 - 3Q_1^2 Q_2^2}{(Q_1^2+m_\pi^2)(Q_2^2+m_\pi^2)}\,,\qquad(5.26)$$

with $\mathbf{Q}_1 = \mathbf{p}_1 - \mathbf{p}_3$ e $\mathbf{Q}_2 = \mathbf{p}_1 - \mathbf{p}_4$.

The first term is associated with the amplitude of the reaction with $1 \to 3$ and $2 \to 4$, whereas the second one corresponds to the the exchange process, in which $1 \to 4$ and $2 \to 3$. Finally, the third contribution takes into account interference between the two amplitudes. Friman and Maxwell also neglect the proton momentum, setting $|\mathbf{Q}_1| = |\mathbf{Q}_2| \approx p_{Fn}$ e $\mathbf{Q}_1 \cdot \mathbf{Q}_2 \approx p_{Fn}^2/2$. Their final result is

$$|M_{fi}^{Mn}|^2 = 16\,G^2\left(\frac{f}{m_\pi}\right)^4 \frac{g_A^2}{E_e^2}\frac{21}{4}\frac{p_{Fn}^4}{(p_{Fn}^2+m_\pi^2)^2}\,.\qquad(5.27)$$

Being independent of the directions of the momenta, the above expression can be moved out of the integral appearing in Eq. (5.21), to obtain $\langle |M_{fi}^{Mn}|^2 \rangle = |M_{fi}^{Mn}|^2$. A comparison between the above result and the exact solution, obtained through numerical integration, shows that the approximation employed in Ref. [100] are remarkably accurate. The discrepancy turns out to be few percent at $n \sim n_0$, and $\sim 10\%$ at $n \sim 3n_0$.

In conclusion, the emissivity associated with the neutron branch of the modified Urca process reported in Ref. [100] turns out to be

$$Q^{Mn} = \frac{11,513}{30,240} \frac{G_F^2 \cos^2 \theta_C \, g_A^2 \, m_n^{*3} \, m_p^*}{2\pi} \left(\frac{f}{m_\pi}\right)^4 \frac{p_{Fp}(k_B T)^8}{\hbar^{10} c^8} p_{Fp} T^8$$

$$\approx 8.1 \times 10^{21} \left(\frac{m_n^*}{m_n}\right)^3 \left(\frac{m_p^*}{m_p}\right) \left(\frac{n_p}{n_0}\right)^{1/3} T_9^8 \, \alpha_n \, \beta_n \ \text{erg cm}^{-3} \, \text{s}^{-1} \, , \qquad (5.28)$$

with $\alpha_n = 1.13$ e $\beta_n = 0.68$.

5.2.2 Proton Branch

Let us now consider the second reaction of (5.17). The initial state proton and neutron will be labelled 1 and 2, respectively, and the final state protons 3 and 4.

The calculation of A turns out to be more complex than in the case of the neutron branch. The result is

$$A_p = \frac{2(2\pi)^5}{p_{Fn} \, p_{Fp}^3 \, p_{Fe}} \left(p_{Fe} + 3p_{Fp} - p_{Fn}\right)^2 \Theta_{Mp} \, , \qquad (5.29)$$

where $\Theta_{Mp} = 1$ if the proton branch is allowed by momentum conservation and $\Theta_{Mp} = 0$ otherwise. Note that, in this case, the triangular translates into the inequality $p_{Fn} < 3p_{Fp} + p_{Fe}$.

The calculation of the squared transition matrix element yields a result identical to Eq. (5.25). Substituting $|\mathbf{Q}_1| = |\mathbf{Q}_2| \approx p_{Fn} - p_{Fp}$ and $\mathbf{Q}_1 \cdot \mathbf{Q}_2 = -(p_{Fn} - p_{Fp})^2$, one obtains an expression similar to Eq. (5.27), except for the replacements $21/4 \to 6$ and $p_{Fn} \to p_{Fn} - p_{Fp}$. Neglecting the angular dependence of the matrix element has been shown to be a good approximation in both the neutron and proton branches.

The proton branch emissivity can be readily obtained from the corresponding neutron branch result using the relation

$$\frac{Q^{Mp}}{Q^{Mn}} = \frac{\langle |M_{fi}^{Mp}|^2 \rangle}{\langle |M_{fi}^{Mn}|^2 \rangle} \left(\frac{m_p^*}{m_n^*}\right)^2 \frac{(p_{Fe} + 3p_{Fp} - p_{Fn})^2}{8 \, p_{Fe} \, p_{Fp}} \Theta_{Mp}$$

$$\approx \left(\frac{m_p^*}{m_n^*}\right)^2 \frac{(p_{Fe} + 3p_{Fp} - p_{Fn})^2}{8 \, p_{Fe} \, p_{Fp}} \Theta_{Mp} \, . \qquad (5.30)$$

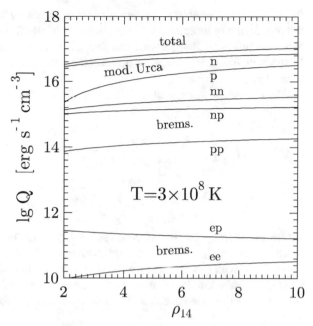

Fig. 5.4 Density dependence of the different mechanisms contributing to the neutrino emissivity of the neutron star core at $T = 3 \times 10^8$ K. The density range is restricted to the region in which the direct Urca process is not allowed. The curves labelled *ep* and *ee* correspond to bremsstrahlung following electron-proton and electron-electron collisions, not discussed in this volume. The matter density is given in units of $10^{14}\,\mathrm{g\,cm^{-3}}$. Reprinted from Ref. [101] with permissions, © Elsevier Science B.V. 2001. All rights reserved

Note that the last line of the above equation is obtained under the commonly employed assumption $\langle |M_{fi}^{Mp}|^2 \rangle = \langle |M_{fi}^{Mn}|^2 \rangle$.

The main difference between the two branches is the presence of a threshold for the proton branch. In *npe* matter, this condition reduces to $p_{Fn} < 4p_{Fp}$, corresponding to a critical proton fraction $x = 1/65 = 0.0154$, which is believed to be attained almost everywhere in the neutron star core. Once the modified Urca process is active in the proton branch, the associated emissivity slowly grows with density, starting from zero and eventually becoming comparable to the emissivity associated with the neutron branch in the vicinity of the threshold of the direct Urca process; see Fig. 5.4.

It has to be pointed out that the temperature dependence of the emissivity of the modified Urca process is described by the power law $Q^{MN} \propto T^8$. The additional factor T^2, with respect to the direct Urca process, originates from the presence of the two additional degenerate fermions.

5.3 Neutrino Bremsstrahlung in Nucleon-Nucleon Collisions

A comprehensive discussion of neutrino emission processes in the neutron star core must include neutrino bremsstrahlung associated with nucleon collisions

$$n + n \to n + n + \nu + \bar{\nu}\,,$$
$$n + p \to n + p + \nu + \bar{\nu}\,, \qquad\qquad (5.31)$$
$$p + p \to p + p + \nu + \bar{\nu}\,.$$

In the above reactions, the scattering process driven by strong interactions leads to the production of a neutrino-antineutrino pair of any flavour. It should be noted that Neutrino bremsstrahlung has no threshold, and is therefore active at any densities. In addition, unlike the processes discussed above, it does not affect the composition of matter.

The emissivity associated with bremsstrahlung can be written in the form

$$Q^{NN} = \int \left[\prod_{j=1}^{4} \frac{d\mathbf{p}_j}{(2\pi)^3} \right] \frac{d\mathbf{p}_\nu}{(2\pi)^3} \frac{d\mathbf{p}'_\nu}{(2\pi)^3} \, \omega_\nu \, (2\pi)^4 \, \delta(E_f - E_i)$$

$$\times \, \delta(\mathbf{P}_f - \mathbf{P}_i) \, f_1 \, f_2 \, (1 - f_3) \, (1 - f_4) \, \frac{1}{s} |M_{fi}|^2 \, , \qquad (5.32)$$

where the index j labels nucleons, $\mathbf{p}_\nu$ and $\mathbf{p}'_\nu$ are the neutrino and antineutrino momenta, $\omega_\nu = E_\nu + E'_\nu$ is the energy carried by the neutrino pair and the symmetry factor $1/s$—with $s = 1$ for the np channel and $s = 4$ for the nn and pp channels—is needed to avoid double counting. In the non relativistic limit, the spin-summed and squared transition matrix element can be written in the form

$$|M_{fi}|^2 = \frac{|\widetilde{M}_{fi}|^2}{\omega_\nu^2} \, , \qquad (5.33)$$

where the denominator ω_ν^2 comes from the propagator of the virtual nucleon appearing in the Feynman diagram of Fig. 5.5.

Neglecting the momenta of the neutrinos with respect to those of the nucleons, $|\widetilde{M}_{fi}|^2$ becomes independent of $\mathbf{p}_\nu$ e $\mathbf{p}'_\nu$. As a consequence, the integration over the phase-space of the neutrinos yields

$$\int_0^\infty E_\nu^2 \, dE_\nu \int_0^\infty E'_\nu \, dE'_\nu \cdots = \frac{1}{30} \int_0^\infty \omega_\nu^5 \, d\omega_\nu \cdots \, , \qquad (5.34)$$

Fig. 5.5 Diagrammatic representation of nucleon-nucleon collisions leading to the emission of a neutrino-antineutrino pair. The dashed and wavy lines represent the nucleon-nucleon and weak interaction, respectively

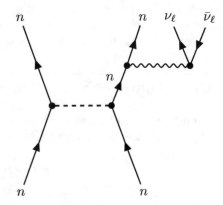

and, using again the decomposition approximation, the emissivity can be cast in the form

$$Q^{NN} = \frac{(2\pi)^4}{(2\pi)^{18}} \frac{1}{30} \, A \, I \, \frac{1}{s} \langle |\tilde{M}_{fi}|^2 \rangle \, T^8 \prod_{j=1}^{4} m_j^* \, p_{Fj} \, , \tag{5.35}$$

with

$$A = (4\pi)^2 \int d\Omega_1 \, d\Omega_2 \, d\Omega_3 \, d\Omega_4 \, \delta(\mathbf{p}_1 + \mathbf{p}_2 - \mathbf{p}_3 - \mathbf{p}_4) \, , \tag{5.36}$$

$$\langle |\tilde{M}_{fi}|^2 \rangle = \frac{(4\pi)^2}{A} \int d\Omega_1 \, d\Omega_2 \, d\Omega_3 \, d\Omega_4 \, \delta(\mathbf{p}_1 + \mathbf{p}_2 - \mathbf{p}_3 - \mathbf{p}_4) \, |\tilde{M}_{fi}|^2 \, , \tag{5.37}$$

$$I = \int_0^\infty dx_\nu \, x_\nu^4 \left[\prod_{j=1}^{4} \int_{-\infty}^{+\infty} dx_j \, f_j \right] \delta\left(\sum_{j=1}^{4} x_j - x_\nu \right) = \frac{164\pi^8}{945} \, . \tag{5.38}$$

Here, $x_j = (E_j - \mu_j)/T$ is the dimensionless nucleon energy, while $x_\nu = \omega_\nu / T$ is the corresponding variable for neutrinos; see Appendix 2. The results of the angular integrations are

$$A_{nn} = \frac{(4\pi)^5}{2 \, p_{Fn}^3} \, , \qquad A_{np} = \frac{(4\pi)^5}{2 \, p_{Fn}^2 \, p_{Fp}} \, , \qquad A_{pp} = \frac{(4\pi)^5}{2 \, p_{Fp}^3} \, . \tag{5.39}$$

As in the case of modified Urca processes, phase-space decomposition provides the rules to rescale different reactions

$$\frac{Q^{np}}{Q^{nn}} = 4 \, \frac{\langle |\tilde{M}_{fi}^{np}|^2 \rangle}{\langle |\tilde{M}_{fi}^{nn}|^2 \rangle} \left(\frac{m_p^*}{m_n^*} \right)^2 \frac{p_{Fp}}{p_{Fn}} \, , \tag{5.40}$$

$$\frac{Q^{pp}}{Q^{nn}} = \frac{\langle |\tilde{M}_{fi}^{pp}|^2 \rangle}{\langle |\tilde{M}_{fi}^{nn}|^2 \rangle} \left(\frac{m_p^*}{m_n^*} \right)^4 \frac{p_{Fp}}{p_{Fn}} \, . \tag{5.41}$$

In the OPE model, the squared matrix element turns out to be

$$|\tilde{M}_{fi}^{NN}|^2 = 16 \, G_F^2 \, g_A^2 \left(\frac{f}{m_\pi} \right)^4 F_{NN} \, , \tag{5.42}$$

with

$$F_{NN} = \frac{\mathbf{Q}_1^4}{(\mathbf{Q}_1^2 + m_\pi^2)^2} + \frac{\mathbf{Q}_2^4}{(\mathbf{Q}_2^2 + m_\pi^2)^2} + \frac{\mathbf{Q}_1^2 \mathbf{Q}_2^2 - 3(\mathbf{Q}_1 \cdot \mathbf{Q}_2)^2}{(\mathbf{Q}_1^2 + m_\pi^2)(\mathbf{Q}_2^2 + m_\pi^2)} \, , \tag{5.43}$$

for nn e pp processes, and

$$F_{NN} = \frac{\mathbf{Q}_1^4}{(\mathbf{Q}_1^2 + m_\pi^2)^2} + \frac{2\mathbf{Q}_2^4}{(\mathbf{Q}_2^2 + m_\pi^2)^2} - 2\,\frac{\mathbf{Q}_1^2\mathbf{Q}_2^2 - (\mathbf{Q}_1 \cdot \mathbf{Q}_2)^2}{(\mathbf{Q}_1^2 + m_\pi^2)(\mathbf{Q}_2^2 + m_\pi^2)}, \qquad (5.44)$$

for the np process. Recall that $\mathbf{Q}_1 = \mathbf{p}_1 - \mathbf{p}_3$, $\mathbf{Q}_2 = \mathbf{p}_1 - \mathbf{p}_4$ and that, in the strong degeneracy limit, $\mathbf{Q}_1 \cdot \mathbf{Q}_2 = 0$.

After averaging the squared matrix element over the directions of the nucleon momenta, one obtains

$$\langle|\tilde{M}_{fi}^{NN}|^2\rangle = 16\,G_F^2\,g_A^2\left(\frac{f}{m_\pi}\right)^4\langle F_{NN}\rangle, \qquad (5.45)$$

where

$$\langle F_{NN}\rangle = 3 - \frac{5}{q}\arctan q + \frac{1}{1+q^2} + \frac{1}{q\sqrt{2+q^2}}\arctan\left(q\sqrt{2+q^2}\right), \qquad (5.46)$$

with $q = 2p_{F_n}/m_\pi$ or $q = 2p_{F_p}/m_\pi$ for nn or pp processes, respectively, and

$$\langle F_{NN}\rangle = 1 - \frac{3\arctan q}{2q} + \frac{1}{2(1+q^2)} + \frac{2\,p_{Fn}^4}{(p_{Fn}^2 + m_\pi^2)^2}$$
$$- \left(1 - \frac{\arctan q}{q}\right)\frac{2\,p_{Fn}^2}{p_{Fn}^2 + m_\pi^2}, \qquad (5.47)$$

with $q = 2p_{F_p}/m_\pi$, for the np process. It should be noted that Eq. (5.47) is obtained under the additional assumption $p_{Fp} \ll p_{Fn}$, which is safely applicable at densities $n \lesssim 3n_0$.

The discussion of neutrino emissivity can be summarised by referring again to the treatment of Friman and Maxwell, and recollecting the approximations involved in their work [100]. In the nn channel they neglect the exchange contribution to the squared matrix element (5.46), average over the directions of the nucleon momenta and set $n = n_0$. Finally, they plug $\langle|\tilde{M}_{fi}|^2\rangle$ obtained within this scheme into the expression of Q^{nn}, with an arbitrary correction factor β_{nn}, meant to take into account all neglected effects, such as correlations and repulsive nuclear forces. The same procedure is applied to the np contribution, in which the interference term of Eq. (5.47) is not taken into account.

The pp process, which is not discussed in Ref. [100], has been studied by Yakovlev and Levenfish [102]. Their results read

$$Q^{nn} = \frac{41}{14,175} \frac{G_F^2 \, g_A^2 \, m_n^{*4}}{2\pi} \left(\frac{f}{m_\pi}\right)^4 p_{Fn} \, \alpha_{nn} \, \beta_{nn} \, T^8 \, \mathcal{N}_\nu$$

$$\approx 7.5 \times 10^{19} \left(\frac{m_n^*}{m_n}\right)^4 \left(\frac{n_n}{n_0}\right)^{1/3} \alpha_{nn} \, \beta_{nn} \, \mathcal{N}_\nu \, T_9^8 \ \mathrm{erg\,cm^{-3}\,s^{-1}} \, , \qquad (5.48)$$

$$Q^{nn} = \frac{82}{14,175} \frac{G_F^2 \, g_A^2 \, m_n^{*2} \, m_p^{*2}}{2\pi} \left(\frac{f}{m_\pi}\right)^4 p_{Fp} \, \alpha_{np} \, \beta_{np} \, (k_B T)^8 \, \mathcal{N}_\nu$$

$$\approx 1.5 \times 10^{20} \left(\frac{m_n^* \, m_p^*}{m_n \, m_p}\right)^2 \left(\frac{n_p}{n_0}\right)^{1/3} \alpha_{np} \, \beta_{np} \, \mathcal{N}_\nu \, T_9^8 \ \mathrm{erg\,cm^{-3}\,s^{-1}} \, ,$$

$$(5.49)$$

$$Q^{pp} = \frac{41}{14,175} \frac{G_F^2 \, g_A^2 \, m_p^{*4}}{2\pi} \left(\frac{f}{m_\pi}\right)^4 p_{Fp} \, \alpha_{pp} \, \beta_{pp} \, (k_B T)^8 \, \mathcal{N}_\nu$$

$$\approx 7.5 \times 10^{19} \left(\frac{m_p^*}{m_p^*}\right)^4 \left(\frac{n_p}{n_0}\right)^{1/3} \alpha_{pp} \, \beta_{pp} \, \mathcal{N}_\nu \, T_9^8 \ \mathrm{erg\,cm^{-3}\,s^{-1}} \, , \qquad (5.50)$$

where $\mathcal{N}_\nu$ is the number of neutrino flavours. The dimensionless factors α_{NN} are obtained from estimates of the matrix elements at $n = n_0$. Their values are $\alpha_{nn} = 0.59$, $\alpha_{np} = 1.06$, and $\alpha_{pp} = 0.11$.

The emissivities of all process turn out to be comparable, with $Q^{pp} < Q^{np} < Q^{nn}$, and $\propto T^8$. This temperature dependence—illustrated in Fig. 5.6 for matter density $\varrho = 2\varrho_0$—can be explained using the usual phase space considerations. The four degenerate nucleons contribute a power T^4 and the two neutrinos a power T^6. The squared matrix element is proportional to ω_ν^{-2}, that is, to T^{-2}, and removes the excess T^2 power.

The results discussed in this chapter are wrapped up in Fig. 5.7, showing the density dependence of the *total* neutrino emissivity of both npe and $npe\mu$ matter at temperature $T = 10^8$, 3×10^8, and 10^9 K. It is apparent that the main effect of the onset of processes involving muons is a shift of the threshold of the direct Urca process towards lower densities.

Fig. 5.6 Temperature
dependence of the different
mechanisms contributing to
the neutrino emissivity of the
neutron star core at $\varrho = 2\varrho_0$.
The curves labelled *ep* and *ee*
correspond to bremsstrahlung
following electron-proton and
electron-electron collisions,
not discussed in this volume.
Reprinted from Ref. [101]
with permissions, © Elsevier
Science B.V. 2001 All rights
reserved

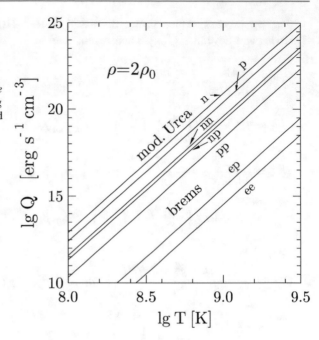

Fig. 5.7 Total neutrino
emissivity in *npe* (solid lines)
and *npeμ* matter (dashed
lines) at $T = 10^8$, 3×10^8,
and 10^9 K, displayed as a
function of matter density
measured in units of
10^{14} g cm^{-3}. Reprinted from
Ref. [101] with permissions,
© Elsevier Science B.V.
2001. All rights reserved

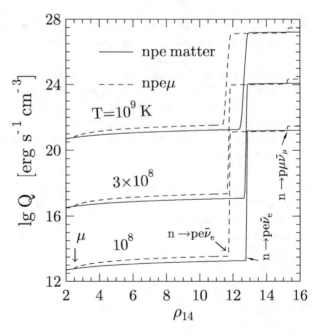

Appendix 1: Neutron β-Decay Rate

Consider the reaction

$$n \longrightarrow p + e + \bar{\nu} \,, \tag{5.51}$$

illustrated by the Feynman diagram of Fig. 5.8, where $p_n \equiv (E_n, \mathbf{p}_n)$, $p_p \equiv (E_p, \mathbf{p}_p)$, $p_e \equiv (E_e, \mathbf{p}_e)$ and $p_\nu \equiv (E_\nu, \mathbf{p}_\nu)$ denote the neutron, proton, electron and antineutrino four-momenta, respectively.

The Fermi interaction Hamiltonian, driving the decay process, is

$$H = \frac{G}{\sqrt{2}} \, j_\mu \ell^\mu \,, \tag{5.52}$$

with $G = G_F \cos\theta_C$, G_F and $\theta_C \sim 13$ deg being the Fermi constant and Cabibbo's angle, respectively. The weak leptonic current ℓ^μ has the $V - A$ form

$$\ell^\mu = \bar{u}_{r'}(p_e)\gamma^\mu \left(1 - \gamma^5\right) v_r(p_\nu) \,, \tag{5.53}$$

where $u_r(p)$ e $v_r(p)$ denote the four-spinors associated with the positive and negative energy solutions of the Dirac equation describing a non interacting fermion of mass m with four-momentum p and spin projection r. They are normalised in such a way as to satisfy the relations

$$\bar{u}_{r'}(p)u_r(p) = \bar{v}_{r'}(p)v_r(p) = 2m \, \delta_{rr'} \,. \tag{5.54}$$

In the non relativistic limit, fully justified in the present context, the hadronic weak current reduces to

$$j_\mu = \chi_{s'}^\dagger \left(\delta_{\mu 0} g_v + \delta_{\mu i} g_A \sigma_i\right) \chi_s \,, \tag{5.55}$$

Fig. 5.8 Feynman diagram illustrating the neutron β-decay process; see Eq. (5.51)

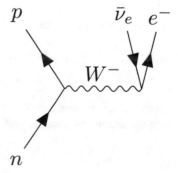

where σ_i (i=1, 2, 3) are Pauli matrices, χ_s and $\chi_{s'}$ are Pauli spinors satisfying the relations $\chi_{s'}^\dagger \chi_s = \delta_{ss'}$ and the values of the vector and axial-vector coupling constants are $g_v = 1$ and $g_A = 1.26$.

The transition rate from the initial state $|i\rangle$ to the set of accessible final states $|f\rangle$ can be obtained using Fermi's golden rule

$$dW_{i\to f} = 2\pi\, \delta(E_n - E_p - E_e - E_v) \sum_{spins} |H_{fi}|^2 \frac{d\mathbf{p}_p}{(2\pi)^3}\frac{d\mathbf{p}_e}{(2\pi)^3}\frac{d\mathbf{p}_v}{(2\pi)^3}, \quad (5.56)$$

where

$$\sum_{spins} |H_{fi}|^2 = (2\pi)^3 \frac{G^2}{2}\delta(\mathbf{p}_n - \mathbf{p}_p - \mathbf{p}_e - \mathbf{p}_v)\frac{1}{2E_e}\frac{1}{2E_v} J_{\lambda\mu} L^{\lambda\mu}, \quad (5.57)$$

with

$$J_{\lambda\mu} = \sum_{spins} j_\lambda^\dagger j_\mu, \quad (5.58)$$

and

$$L^{\lambda\mu} = \sum_{spins} \ell^{\lambda\dagger}\ell^\mu. \quad (5.59)$$

Substituting the leptonic weak current in Eq. (5.59) and exploiting the completeness relations fulfilled by Dirac's spinors, one finds

$$L^{\lambda\mu} = \sum_{rr'} \bar{u}_{r'}(p_e)\gamma^\lambda(1-\gamma^5)v_r(p_v)\bar{v}_r(p_v)\gamma^\mu(1-\gamma^5)u_{r'}(p_e)$$

$$= 8\left[p_e^\lambda p_v^\mu + p_e^\mu p_v^\lambda - g^{\lambda\mu}(p_e p_v) + i\epsilon^{\lambda\sigma\mu\rho}p_{e_\sigma}p_{v_\rho}\right], \quad (5.60)$$

where $g^{\lambda\mu} = \mathrm{diag}(1,-1,-1,-1)$ and $\epsilon^{\lambda\sigma\mu\rho}$ are the metric tensor and the fully antisymmetric unit tensor.

The explicit expression of $J_{\lambda\mu}$ is obtained from Eq. (5.55), implying

$$j^0 = \chi_{s'}^\dagger \chi_s = \delta_{ss'}, \quad j^i = g_A\chi_{s'}^\dagger\sigma^i\chi_s, \quad i = 1, 2, 3. \quad (5.61)$$

Carrying out the spin sum one readily finds that the tensor $J_{\lambda\mu}$ is diagonal, with its non vanishing components being given by

$$J^{00} = 2, \quad J^{11} = J^{22} = J^{33} = 2g_A^2. \quad (5.62)$$

Finally, substitution of Eqs. (5.60) and (5.62) in the expression of the differential decay rate leads to

$$dW_{i \to f} = (2\pi)^4 \delta(E_n - E_p - E_e - E_\nu)\delta(\mathbf{p}_n - \mathbf{p}_p - \mathbf{p}_e - \mathbf{p}_\nu)$$
$$\times G^2 \left[1 + \cos\theta + g_A{}^2(3 - \cos\theta)\right] \frac{d\mathbf{p}_p}{(2\pi)^3} \frac{d\mathbf{p}_e}{(2\pi)^3} \frac{d\mathbf{p}_\nu}{(2\pi)^3} , \quad (5.63)$$

where θ is the angle between the lepton momenta $\mathbf{p}_e$ and $\mathbf{p}_\nu$.

Neglecting the contribution of $\mathbf{p}_\nu$ in the argument of the δ-function expressing momentum conservation—which is a remarkably accurate approximation in the case of emission of thermal antineutrinos in dense degenerate matter—we can perform the $\cos\theta$ integration, yielding the final result

$$dW_{i \to f} = 2\pi \, \delta(E_n - E_p - E_e - E_\nu)\delta(\mathbf{p}_n - \mathbf{p}_p - \mathbf{p}_e)$$
$$\times 2G^2 \left(1 + 3g_A{}^2\right) 4\pi E_\nu^2 \, dE_\nu \frac{d\mathbf{p}_p}{(2\pi)^3} \frac{d\mathbf{p}_e}{(2\pi)^3} . \quad (5.64)$$

Appendix 2: Rate of the Direct Urca Process

This appendix describes the calculation of the neutrino emission rate for the nucleonic Urca process, performed using the method based on phase space decomposition.

Owing to the strong degeneracy of nucleons and electrons, the main contributions to the integral appearing in the definition of the emissivity

$$Q^D = 2 \int \frac{d\mathbf{p}_n}{(2\pi)^3} dW_{i \to f} \, f_n(1 - f_p)(1 - f_e), \quad (5.65)$$

originate from momenta lying within a thin shell surrounding the Fermi surface. As a consequence, one can set $|\mathbf{p}_i| = p_{F_i}$ in the integrand, and assume that the typical energy exchanged in the reactions is $\sim T$.

Because the energy carried by the neutrino is $E_\nu \sim T$, the corresponding momentum, being of the same order, is much smaller than the momenta of the degenerate fermions involved in the process. Therefore, it can be safely neglected in the argument of the momentum-conserving δ-function. The resulting expression of the transition rate is given by

$$dW_{i \to f} \frac{d\mathbf{P}_n}{(2\pi)^3} = \frac{(2\pi)^4}{(2\pi)^{12}} \delta(E_n - E_p - E_e - E_\nu) \, \delta(\mathbf{p}_n - \mathbf{p}_p - \mathbf{p}_e)$$
$$\times |M_{fi}|^2 \, 4\pi E_\nu^2 dE_\nu \prod_{j=1}^{3} p_{Fj} m_j^* dE_j d\Omega_j , \quad (5.66)$$

where $d\Omega_j$ is the differential solid angle specifying the direction of the momentum $\mathbf{p}_j$, $m_j^* = p_{Fj}/v_{Fj}$ is the effective mass of the particles of species j and $v_{Fj} = (\partial E_j/\partial p_j)_{p=p_{Fj}}$, denotes the corresponding Fermi velocity.

It is convenient to introduce the new dimensionless variables

$$x_\nu = \frac{E_\nu}{T} \quad , \quad x_j = \frac{E_j - \mu_j}{T} \,,$$

with $j = n$, p, e. Replacing $x_j \rightarrow -x_j$ for $j = p$, e, allows one to exploit the property of the Fermi-Dirac distribution $1 - f(x_j) = f(-x_j)$. In addition, the result of the angular integration can be written in terms of the quantity

$$A = 4\pi \int d\Omega_1 d\Omega_2 d\Omega_3 \,. \tag{5.67}$$

Using the above definitions, Eq. (5.65) can be rewritten in the form

$$Q^D = \frac{2}{(2\pi)^8} |M_{fi}|^2 A \left[T^7 \int_0^\infty dx_\nu x_\nu^3 \left(\int_{-\frac{\mu_1}{T}}^\infty \int_{-\infty}^{\frac{\mu_2}{T}} \int_{-\infty}^{\frac{\mu_3}{T}} dx_j f_j \right) \right.$$
$$\left. \times \frac{1}{T} \delta(x_1 + x_2 + x_3 - x_\nu) \right] \prod_{j=1}^{3} p_{Fj} m_j^* \,. \tag{5.68}$$

Because the system is strongly degenerate, one can replace $\mu_j/T \rightarrow \infty$, the associated error being exponentially small. Thus, one finally obtains

$$Q^D = \frac{2}{(2\pi)^8} |M_{fi}|^2 T^6 A I \prod_{j=1}^{3} p_{Fj} m_j^* \,, \tag{5.69}$$

where

$$I = \int_0^\infty dx_\nu x_\nu^3 \left[\prod_{j=1}^{3} \int_{-\infty}^\infty dx_j f_j \right] \delta(x_1 + x_2 + x_3 - x_\nu) \,. \tag{5.70}$$

Calculation of A

The momentum conserving δ-function appearing in Eq. (5.67) can be rewritten in spherical coordinates using the property

$$\int \delta_s(\mathbf{a} - \mathbf{b}) \, d\mathbf{a} = \int \delta(a_x - b_x) \, da_x \int \delta(a_y - b_y) \, da_y \int \delta(a_z - b_z) \, da_z = 1 \,.$$

The requirement that the above relation still hold true in spherical coordinates, that is, that

$$\int \delta_s(\mathbf{a} - \mathbf{b})\, da = \int \delta_s(\mathbf{a} - \mathbf{b})\, a^2 da\, d\Omega_a = 1\,,$$

implies

$$\delta_s(\mathbf{a} - \mathbf{b}) = \delta(a - b)\,\frac{\delta(\Omega_a - \Omega_b)}{a^2}\,.$$

In the case under consideration the above expression becomes

$$A = 4\pi \int d\Omega_1 d\Omega_2 d\Omega_3\, \delta(p_n - |\mathbf{p}_p + \mathbf{p}_e|)\,\frac{\delta(\Omega_1 - \Omega_{2+3})}{p_n^2}$$

$$= 4\pi \int d\Omega_2 d\Omega_3\, \frac{\delta(p_n - |\mathbf{p}_p + \mathbf{p}_e|)}{p_n^2}\,. \tag{5.71}$$

The radial δ-function can be transformed into a δ-function in one of the angular variables. One first rewrites the δ-function in the form

$$\delta[f(\cos\theta_2)] = \delta[p_n - (p_p^2 + p_e^2 + 2p_p p_e \cos\theta_2)^{\frac{1}{2}}]\,,$$

with the z-axis chosen along the direction of the electron momentum $\mathbf{p}_e$. The root of the equation $f(\cos\theta_2) = 0$

$$\cos\theta_2 = \frac{p_n^2 - p_p^2 - p_e^2}{2 p_p p_e}\,,$$

and the result

$$|f'(\overline{\cos\theta_2})| = \frac{p_p p_e}{p_n}\,,$$

can be used to obtain

$$\frac{\delta(p_n - |\mathbf{p}_p + \mathbf{p}_e|)}{p_n^2} = \frac{1}{p_n^2}\,\frac{1}{|f'(\overline{\cos\theta_2})|}\,\delta(\cos\theta_2 - \overline{\cos\theta_2})$$

$$= \frac{1}{p_n p_p p_e}\,\delta(\cos\theta_2 - \overline{\cos\theta_2})\,,$$

Finally, replacing the above equation into (5.71) leads to

$$A = \frac{32\pi^3}{p_{Fn} p_{Fp} p_{Fe}}\,\Theta_{npe}\,, \tag{5.72}$$

where the function Θ_{npe} takes into account the triangular condition on the Fermi momenta of the degenerate fermions; see Eq. (5.12).

Calculation of I

The quantity I, defined by Eq. (5.70), can be conveniently rewritten in the form

$$I = \int_0^\infty dx_\nu x_\nu^3 \, J(x_\nu) \, , \tag{5.73}$$

with

$$J(x_\nu) = \int_{-\infty}^\infty \prod_{j=1}^3 dx_j (1 + e^{x_j})^{-1} \, \delta \left(\sum_{j=1}^3 x_j - x_\nu \right) . \tag{5.74}$$

The calculation of J is based on the substitution of the δ-function with its definition in terms of the Fourier transform

$$\delta(x) = \frac{1}{2\pi} \int_{-\infty}^\infty dz \, e^{izx} \, ,$$

yielding

$$
\begin{aligned}
J(x_\nu) &= \frac{1}{2\pi} \int_{-\infty}^\infty dz \int_{-\infty}^\infty \prod_{j=1}^3 dx_j (1 + e^{x_j})^{-1} e^{iz(x_j - x_\nu)} \\
&= \frac{1}{2\pi} \int_{-\infty}^\infty dz \, e^{-izx_\nu} \left(\int_{-\infty}^\infty dx \, (1 + e^x)^{-1} e^{izx} \right)^3 \\
&= \frac{1}{2\pi} \int_{-\infty}^\infty dz \, e^{-izx_\nu} \, [f(z)]^3 \, ,
\end{aligned}
\tag{5.75}
$$

where

$$f(z) = \int_{-\infty}^\infty dx \, (1 + e^x)^{-1} e^{izx} \, . \tag{5.76}$$

To obtain the expression of $f(z)$, consider the integral

$$K = \oint dx \, (1 + e^x)^{-1} e^{izx} \, , \tag{5.77}$$

to be evaluated using Cauchy's theorem with the contour shown in Fig. 5.9.

Fig. 5.9 Integration contour employed for the evaluation of K; see Eq. (5.77)

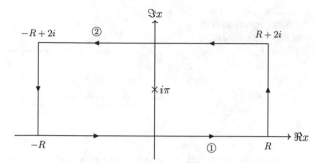

As $R \to \infty$ the contributions of the vertical sides become negligible. Along the real axis, marked with ①, $K = f(z)$, whereas along the path marked with ②

$$K = \int_{-\infty}^{\infty} dx \, (1 + e^{\Re x + 2i\pi})^{-1} e^{iz(\Re x + 2i\pi)} = e^{-2\pi z} f(z) \, .$$

It follows that

$$K = f(z) - e^{-2\pi z} f(z) = (1 - e^{-2\pi z}) f(z) \, .$$

The value of K can be obtained using the method of residues. From

$$\text{Res} \, K(z) = \lim_{x \to i\pi} (x - i\pi) \frac{e^{izx}}{1 + e^x} = \lim_{x \to i\pi} (x - i\pi) \frac{e^{izx}}{1 - \sum_{n=0}^{\infty} \frac{1}{n!}(x - i\pi)^n}$$

$$= \lim_{x \to i\pi} -\frac{e^{izx}}{\sum_{n=1}^{\infty} \frac{1}{n!}(x - i\pi)^{(n-1)}} = -e^{-\pi z} \, ,$$

it follows that

$$K(z) = (1 - e^{-2\pi z}) f(z) = -2i\pi e^{-\pi z} \, ,$$

and using the above result one finally finds

$$f(z) = \frac{\pi}{i \sinh \pi z} \, . \tag{5.78}$$

The above results implies that J can be written in the form

$$J(x_v) = \frac{1}{2i\pi} \int_{-\infty - i\epsilon}^{\infty - i\epsilon} dz \, e^{-izx_v} \left(\frac{\pi}{\sinh \pi z} \right)^3 \, , \tag{5.79}$$

where the quantity $-i\epsilon$, with $\epsilon = 0^+$ takes into account the presence of a pole of third order in the integrand. To determine a suitable integration contour, it is

Fig. 5.10 Integration contour employed for the evaluation the integral of Eq. (5.81)

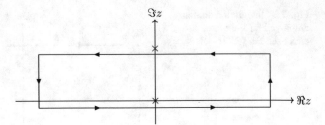

convenient to carry out the transformation $z = z' - i$, leading to

$$J(x_\nu) = \frac{e^{-x_\nu}}{2i\pi} \int_{-\infty-i\epsilon+i}^{\infty-i\epsilon+i} dz\, e^{-izx_\nu} \left(\frac{\pi}{\sinh \pi z}\right)^3 . \tag{5.80}$$

By summing Eqs. (5.79) and (5.80), one obtains the result of the integration performed along the contour shown in Fig. 5.10

$$(1 + e^{x_\nu})\, J(x_\nu) = \frac{1}{2i\pi} \left[\int_{-\infty-i\epsilon}^{\infty-i\epsilon} + \int_{-\infty-i\epsilon+i}^{\infty-i\epsilon+i} \right] dz\, e^{-izx_\nu} \left(\frac{\pi}{\sinh \pi z}\right)^3$$

$$= \frac{1}{2i\pi} \oint dz\, e^{-izx_\nu} \left(\frac{\pi}{\sinh \pi z}\right)^3 . \tag{5.81}$$

The above equation implies that

$$(1 + e^{x_\nu})\, J = -\mathrm{Res}\left[e^{-izx_\nu} \left(\frac{\pi}{\sinh \pi z}\right)^3 \right]_{z=0}$$

$$= -\lim_{z \to 0} \frac{1}{2} \frac{d^2}{dz^2}\left[z^3 e^{-izx_\nu} \left(\frac{\pi}{\sinh \pi z}\right)^3 \right]$$

$$= -\lim_{z \to 0} \frac{1}{2} \left\{ -x_\nu^2 e^{-izx_\nu} \left(\frac{\pi}{\sinh \pi z}\right)^3 \right.$$

$$+ 6i x_\nu e^{-izx_\nu} \left(\frac{\pi}{\sinh \pi z}\right)^3 \left(\pi z^3 \frac{\cosh \pi z}{\sinh \pi z} - z^2\right)$$

$$+ \left[6z - 18\pi z^2 \frac{\cosh \pi z}{\sinh \pi z} + 3\pi^2 z^3 \frac{1}{(\sinh \pi z)^2} + 9\pi^2 z^3 \left(\frac{\cosh \pi z}{\sinh \pi z}\right)^2 \right]$$

$$\left. \times \frac{\pi^3 e^{-izx_\nu}}{(\sinh \pi z)^3} \right\}$$

In the $z \to 0$ limit, the second contribution to the first line of the above result vanishes, as can be easily seen by substituting the first order expansion

$$\frac{\cosh \pi z}{\sinh \pi z} \approx \frac{1}{\pi z} ,$$

while the first contribution reduces to $-x_\nu^2$. The second line must be expanded up to the zero-th order term. The resulting contributions are

① : $\quad \dfrac{6\pi^3 z}{(\pi z)^3 \left[1 + \frac{(\pi z)^2}{2}\right]} \approx \dfrac{6}{z^2} \left(1 - \dfrac{1}{2}\pi^2 z^2\right) = \dfrac{6}{z^2} - 3\pi^2 ,$

② : $\quad \dfrac{3\pi^5 z^3}{(\pi z)^5 \left[1 + \frac{5(\pi z)^2}{6}\right]} \approx \dfrac{3}{z^2} - \dfrac{5}{2}\pi^2 ,$

③ : $\quad \dfrac{9\pi^5 z^3}{(\pi z)^5 \left[1 + \frac{5(\pi z)^2}{6}\right]} (1 + \pi^2 z^2) \approx \dfrac{9}{z^2} + \dfrac{3}{2}\pi^2 .$

④ : $\quad -\dfrac{18\pi^4 z^2}{(\pi z)^4 \left[1 + \frac{2(\pi z)^2}{3}\right]} \left(1 + \dfrac{1}{2}\pi^2 z^2\right) \approx -\dfrac{18}{z^2} + 3\pi^2 ,$

Summing the above contributions one arrives at

$$(1 + e^{x_\nu}) J(x_\nu) = \frac{x_\nu^2 + \pi^2}{2} ,$$

implying

$$J(x_\nu) = \frac{\pi^2 + x_\nu^2}{2(1 + e^{x_\nu})} . \tag{5.82}$$

The integral

$$I = \int_0^\infty dx_\nu x_\nu^3 J(x_\nu) ,$$

can be performed using Eq. (3.411) of Ref. [103]

$$I(p, n) = \int_0^\infty \frac{x^{2n-1}}{e^{px} + 1} dx = (1 - 2^{1-2n}) \left(\frac{2\pi}{p}\right)^{2n} \frac{|B_{2n}|}{4n} , \tag{5.83}$$

where B_n is a Bernoulli number. In the case under consideration, involving $p = 1$ and $n = 2, 3$, one obtains

$$I = \frac{457\pi^6}{5040}.$$ (5.84)

Finally, substitution into Eq. (5.8), yields the expression of the neutrino emissivity of Eq. (5.13).

Neutron Star Structure and Dynamics

6

Abstract

The surface gravity of a neutron star of mass M and radius R—given by GM/R, with G being the gravitational constant—exceeds the value typical of white dwarfs by about three orders of magnitude. Owing to the strength of the gravitational field, the effects predicted by Einstein's theory of general relativity play a critical role in determining the neutron star structure and dynamics, and must be properly taken into account. This chapter provides an outline of the derivation of neutron star properties relevant to astrophysical observation—including mass, radius, cooling rate, and tidal deformability—based on the relativistic formalism.

6.1 Hydrostatic Equilibrium

The effects of space-time distortion turn out to be negligible when the surface gravitational potential of a star of mass M and radius R fulfills the requirement $GM/R \ll 1$, with G being the gravitational constant. This condition is largely satisfied by white dwarfs, having $GM/R \sim 10^{-4}$, but not by neutron stars, whose higher density entails much higher values of surface gravity, typically of order 10^{-1}.

The Newtonian formalism employed for the description of white dwarfs provides an accurate approximation to the equation of stellar equilibrium when matter density is not as large as to induce a significant space-time curvature. In this case, the metric can be written as

$$ds^2 = \eta_{\mu\nu}dx^\mu dx^\nu , \tag{6.1}$$

© The Author(s), under exclusive license to Springer Nature Switzerland AG 2023
O. Benhar, *Structure and Dynamics of Compact Stars*, Lecture Notes
in Physics 1019, https://doi.org/10.1007/978-3-031-35628-5_6

with

$$\eta_{\mu\nu} = \text{diag}\,(-1, 1, 1, 1).\tag{6.2}$$

In Einstein's theory of general relativity Eq. (6.1) is replaced by

$$ds^2 = g_{\mu\nu}dx^\mu dx^\nu\,,\tag{6.3}$$

with the metric tensor $g_{\mu\nu}$ being a function of space-time coordinates.

Relativistic corrections to the hydrostatic equilibrium equations (1.52) and (1.53) are obtained from the solution of Einstein's field equations

$$G_{\mu\nu} = 8\pi G T_{\mu\nu}\,,\tag{6.4}$$

where G is the gravitational constant, $T_{\mu\nu}$ is the energy-momentum tensor, and the Einstein's tensor $G_{\mu\nu}$ is defined in terms of $g_{\mu\nu}$, encoding space-time geometry. Equations (6.4) state the relation between the distribution of matter, described by $T_{\mu\nu}$, and space-time curvature, described by $g_{\mu\nu}$.

Consider a star consisting of a static and spherically symmetric distribution of matter in chemical, hydrostatic and thermodynamic equilibrium. The metric of the corresponding gravitational field reduces to the form

$$ds^2 = g_{\mu\nu}dx^\mu dx^\nu = -e^{2\Phi(r)}dt^2 + e^{2\lambda(r)}dr^2 + r^2(d\theta^2 + \sin^2\theta d\varphi^2)\,,\tag{6.5}$$

implying[1]

$$g_{\mu\nu} = \text{diag}\left(-e^{2\Phi(r)}, e^{2\lambda(r)}, r^2, r^2\sin^2\theta\right),\tag{6.6}$$

$v(r)$ e $\lambda(r)$ being functions to be determined solving Einstein's equations; see, e.g., Ref. [104].

Under the assumption that matter in the star interior behave as an ideal fluid, the energy-momentum tensor reduces to the form

$$T_{\mu\nu} = (\epsilon + P)u_\mu u_\nu + P g_{\mu\nu}\,,\tag{6.7}$$

where ϵ and P denote energy density and pressure, respectively, and

$$u_\mu = \frac{dx_\mu}{d\tau} \equiv (e^{-\Phi(r)}, 0, 0, 0)\,,\tag{6.8}$$

---

[1] Here, and in the following, $x^0 = t$, $x^1 = r$, $x^2 = \theta$, $x^3 = \varphi$.

is the local four-velocity of the fluid at rest. From the above equations, it follows that

$$T_{\mu\nu} = \text{diag}\left(\epsilon e^{2\Phi(r)}, Pe^{2\lambda(r)}, Pr^2, Pr^2 \sin^2\theta\right) \tag{6.9}$$

The Einstein tensor $G_{\mu\nu}$, given by

$$G_{\mu\nu} = R_{\mu\nu} - \frac{1}{2}g_{\mu\nu}R \,, \tag{6.10}$$

with

$$R = g^{\mu\nu}R_{\mu\nu} \,, \tag{6.11}$$

depends on $g_{\mu\nu}$ through the Christoffel symbols

$$\Gamma^{\lambda}{}_{\mu\nu} = \frac{1}{2}g^{\lambda\alpha}(g_{\alpha\mu,\nu} + g_{\alpha\nu,\mu} - g_{\mu\nu,\alpha}) \,, \tag{6.12}$$

which in turn appear in the definition of the Ricci tensor

$$R_{\mu\nu} = \Gamma^{\alpha}{}_{\mu\nu,\alpha} - \Gamma^{\alpha}{}_{\mu\alpha,\nu} - \Gamma^{\alpha}{}_{\beta\nu}\Gamma^{\beta}{}_{\mu\alpha} + \Gamma^{\alpha}{}_{\beta\alpha}\Gamma^{\beta}{}_{\mu\nu} \,. \tag{6.13}$$

Note that the above equations are written using the standard concise notation $g_{\alpha\mu,\nu} = \partial g_{\alpha\mu}/\partial x^{\nu}$. The quantity R defined by Eq. (6.11) is referred to as Ricci curvature, or scalar curvature.

From Eqs. (6.6), (6.12) and (6.13) it follows that the only nonvanishing elements of $R_{\mu\nu}$ are

$$R_{00} = \left[\Phi'' + (\Phi')^2 - \lambda'\Phi' + \frac{2\Phi'}{r}\right]e^{2(\Phi-\lambda)} \,,$$

$$R_{11} = \lambda'\Phi' - \Phi'' - (\Phi')^2 + \frac{2\lambda'}{r} \,,$$

$$R_{22} = \left[r(\lambda' - \Phi') - 1\right]e^{-2\lambda} + 1 \,,$$

$$R_{33} = R_{22}\sin^2\theta \,, \tag{6.14}$$

with $\Phi' = d\Phi/dr$, implying

$$R = 2e^{-2\lambda}\left[\Phi'' + (\Phi')^2 - \lambda'\Phi' + \frac{2}{r}(\Phi' - \lambda') + \frac{1}{r^2}(1 - e^{2\lambda})\right] \,. \tag{6.15}$$

Substitution of Eqs. (6.14) and (6.15) into Eqs. (6.4) and (6.10) yields

$$G_{00} = \frac{1}{r^2} e^{2\Phi} \frac{d}{dr} \left[r(1 - e^{-2\lambda}) \right] = 8\pi G \, T_{00} \, , \tag{6.16}$$

$$G_{11} = -\frac{1}{r^2} e^{2\lambda} (1 - e^{-2\lambda}) + \frac{2}{r} \Phi' = 8\pi G \, T_{11} \, , \tag{6.17}$$

$$G_{22} = r^2 e^{-2\lambda} \left[\Phi'' + (\Phi')^2 + \frac{\Phi'}{r} - \Phi' \lambda' - \frac{\lambda'}{r} \right] = 8\pi G \, T_{22} \, , \tag{6.18}$$

$$G_{33} = G_{22} \sin^2 \theta = 8\pi G \, T_{33} \, . \tag{6.19}$$

Note that the last equation, involving G_{33} and T_{33}, does not provide any additional information, because $T_{33} = T_{22} \sin^2 \theta$; see Eq.(6.9).

Defining now the function $M(r)$ such that

$$\frac{dM}{dr} = \frac{1}{2} r(1 - e^{-2\lambda}) \, , \tag{6.20}$$

one can rewrite Eq.(6.16) in the form

$$\frac{dM}{dr} = 4\pi r^2 \epsilon(r) \, , \tag{6.21}$$

implying

$$M(r) = 4\pi \int_0^r dr' r'^2 \epsilon(r'), \tag{6.22}$$

The above equation clearly shows the analogy between $M(r)$ and the mass of equilibrium configurations in Newton's theory; see Eq. (1.53).

Equation (6.17) can also be rewritten using Eq. (6.21), implying

$$e^{-2\lambda} = 1 - 2 \frac{GM(r)}{r} \, . \tag{6.23}$$

The resulting expression is

$$\Phi' = \frac{M + 4\pi r^3 P(r)}{r^2} \left(1 - 2\frac{GM}{r} \right)^{-1} \, . \tag{6.24}$$

Combining the above result with the equation

$$\Phi' = -\frac{2}{\epsilon(r) + P(r)} \frac{dP}{dr} \, , \tag{6.25}$$

following from the conservation equation $\nabla^\nu T_{\mu\nu} = 0$—with ∇^ν being the covariant divergence—one can finally write the equilibrium equations of general relativity in the form originally derived by Tolman, Oppenheimer and Volkoff (TOV) [17, 18]

$$\frac{dP(r)}{dr} = -\epsilon(r)\frac{GM(r)}{r^2}\left[1 + \frac{P(r)}{\epsilon(r)}\right]\left[1 + \frac{4\pi r^3 P(r)}{M(r)}\right]\left[1 - \frac{2GM(r)}{r}\right]^{-1}$$

(6.26)

with $M(r)$ defined by Eq. (6.22).

The first factor of the product appearing in the right hand-side of Eq. (6.26) has the same form as the newtonian gravitational force in Eq. (1.52), but with matter density replaced by energy density. Note that, because $\epsilon(r)$ also determines the mass through Eq. (6.22), the resulting $M(r)$ turns out to be the sum of rest mass, internal energy and gravitational energy. The first two additional factors in square brackets take into account relativistic corrections that become vanishingly small in the limit $p_F/m \to 0$, m and p_F being the mass and Fermi momentum of the matter constituents, respectively. Finally, the third factor describes the effect of space-time curvature. In the non relativistic limit $\epsilon(r) \to \varrho(r)$, all three factors reduce to unity, and the classical equilibrium equation—that is, Eq. (1.52)—is recovered.

The TOV equations describe the hydrostatic equilibrium between matter pressure and gravitational pull in general relativity. Being two first-order differential equations, they can be solved assigning the values of $P(r)$ and $M(r)$ at, e.g., the center of the star, corresponding to $r = 0$. It follows that, because $M(0) = 0$, the solutions are specified by the single parameter $\epsilon_c = \epsilon(0)$, the central energy density of the star, which determines $P(0)$ through the EOS $P(\epsilon)$. For any given EOS the TOV equations are integrated numerically from the center of the star to its surface, corresponding to $r = R$—with the radius R being dictated by the condition $P(R) = 0$—and the mass of the star is given by $M(R)$.

It has to be kept in mind, however, that solving the TOV equations with the appropriate boundary conditions allows to determine all equilibrium configurations, but equilibrium does not necessarily entalis stability. This point is illustrated in Fig. 6.1, showing the typical dependence of the neutron star mass on central energy density.

The maximum mass, analog to the Chandrasekhar mass of white dwarfs, corresponds to the energy density $\bar{\epsilon}_c$ such that

$$\left(\frac{dM}{d\epsilon_c}\right)_{\epsilon_c=\bar{\epsilon}_c} = 0,$$

(6.27)

and the stable configurations are those in the region of $dM/d\epsilon_c > 0$. In the region in which $dM/d\epsilon_c < 0$, on the other hand, equilibrium configurations turn out to be unstable with respect to radial oscillations of the star.

The sensitivity of the mass-radius relation to the composition and dynamics of neutron star matter is illustrated in Fig. 6.2, taken from Ref. [105]. The most striking feature is the effect of the appearance of hyperons, leading to a large decrease of the

Fig. 6.1 Neutron star mass, obtained from the solution of the TOV equations using the nuclear matter EOS of Ref. [63], displayed as a function of the central energy density

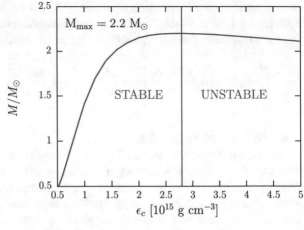

Fig. 6.2 Mass-radius relation obtained from different models of neutron star matter. The lines labelled 1 and 2 correspond to nucleon matter, while lines 3 and 4 have been obtained including hyperons. See text for details. Adapted from Ref. [105] with permissions, © EPLA 2011. All rights reserved

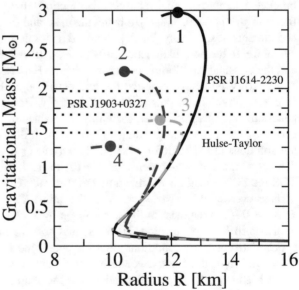

maximum mass M_{max}. The results obtained using two EOSs of pure nuclear matter, characterised by compressibility modules $K = 285$ and 211 MeV, are represented by the curves labelled 1 and 2, respectively, while lines 3 and 4 show the mass-radius relations corresponding to EOSs of matter consisting of nucleons, Λs and Σs. The difference between the two curves originates from the inclusion of irreducible three-body interactions involving nucleons and hyperons, leading to an increase of M_{max}. The resulting maximum is compatible with the canonical neutron star mass obtained from astronomical observations, $M \gtrsim 1.4\ M_\odot$, but falls short of explaining the recently measured masses of millisecond pulsars, whose values turn out to be around and above $2\ M_\odot$. The *hyperon puzzle* will be discussed further in Chap. 7.

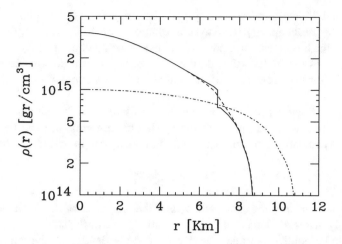

Fig. 6.3 Density profiles of a neutron star of mass 1.4 $M_\odot$ described by different models. The dot-dash line has been obtained using the purely nucleonic APR EOS, whereas the solid and dashed lines correspond to hybrid stars in which the transition to quark matter is associated with a sharp interface or to the formation of a mixed phase. Reprinted from Ref. [89] with permissions, © ESO 2005. All rights reserved

Solving the TOV equations also gives access to the density profile of stable neutron stars, which turns out to be strongly affected by the occurrence of phase transitions in the neutron star core. As an example, Fig. 6.3 shows a comparison between the density of a neutron star of mass 1.4 $M_\odot$ described by the APR EOS, represented by the dot-dash line, and a *hybrid* star of the same mass, in which the nuclear matter phase is combined with a core of quark matter described using the MIT bag model. The solid and dashed lines have been obtained assuming a sharp transition from nuclear to quark matter or the formation of a mixed phase, respectively.

6.2 Cooling

The dominant cooling mechanism in the first stage of the neutron star thermal evolution is neutrino emission, which lowers the temperature from the initial value $T \sim 10^{11}$ to $\sim 10^8$ K through the processes discussed in Chap. 5. The second important factor driving the time-dependence of the temperature is heat transport from the inner region of the star to the surface, driving the observed thermal photon emission.

The equation describing neutron star cooling can be written in a somewhat simplified form as

$$\frac{dE}{dt} = C_V \frac{dT}{dt} = -L_\nu - L_\gamma \,, \tag{6.28}$$

where E and T denote internal energy and temperature, respectively, while $C_V = dE/dT$ is the heat capacity. The two contributions appearing in the right-hand side account for the luminosity originating from neutrino and photon emission. Note that Eq. (6.28) does not include the additional contribution of a positive *heating* term, associated with the occurrence of dissipative processes converting internal energy into heat.

The rate of energy loss from neutrino emission can be obtained from the emissivities derived in Chap. 5. Assuming that photon emission can be described as blackbody radiation, the corresponding luminosity can be written in the form

$$L_\gamma = 4\pi R^2 \sigma T_s^4 , \tag{6.29}$$

where R and σ are the star radius and the constant of Stephan-Boltzmann, respectively, while T_s denotes the *surface* temperature of the star.

It has to be pointed out that, because L_ν and C_V depend on the temperature of the star interior, while L_γ is a function of surface temperature, the determination of the cooling rate requires an additional equation, describing the relationship between T and T_s. The following section provides a brief outline of the treatment of heat transport and thermal equilibrium within the formalism of general relativity; a detailed derivation can be found in Ref. [106]. Note that this discussion is based on the tenet that the neutron star equilibrium configuration can be obtained from the solution of the TOV equations using a zero-temperature EOS.

Thermal Evolution in General Relativity
The equation describing energy transport in neutron stars is obtained combining the contributions of radiative and conductive transfer. Compared to these mechanisms, convective transfer turns out to be negligible, because the high thermal conductivity of the degenerate fermion gases prevents the temperature gradients from reaching the adiabatic limit corresponding to the onset of convection. The resulting expression is

$$\frac{\partial}{\partial r}(e^\Phi T) = -\frac{e^\Phi L_r}{4\pi \kappa r^2}\left(1 - 2\frac{GM}{r}\right)^{-1/2} , \tag{6.30}$$

where L_r is the *local luminosity*, defined as the non-neutrino heat flux transported through a sphere of radius r, and κ is the thermal conductivity. The gravitational redshift e^Φ is a function of the of the gravitational potential Φ defined by Eq. (6.25), which is in turn determined by the dynamics underlying the EOS.

The equation of thermal equilibrium describes the balance of all energy conversion processes taking place in the star interior. These include:

• conversion of rest mass into internal energy through nuclear reactions;
• work done by the gravitational force;
• energy loss though radiation, conduction and convection.

Fig. 6.4 Thermal evolution of two neutron stars described by the EOS of Ref. [63]. The curve labelled 1.4 $M_\odot$ corresponds to the slow, or *standard* cooling scenario, in which the direct Urca mechanism is not allowed, while the fast, or *enhanced* cooling scenario is illustrated by the curve labelled 1.9 $M_\odot$

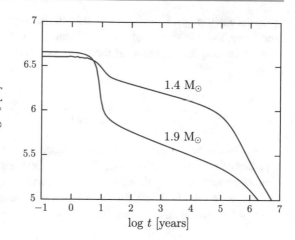

The relativistic counterpart of the equilibrium equation (6.28) can be written in the form

$$\frac{\partial}{\partial t}(e^\Phi T) = -\frac{1}{C_V}\left[e^{2\Phi}Q_\nu + \frac{1}{4\pi r^2}\left(1 - 2\frac{GM}{r}\right)^{1/2}\frac{\partial}{\partial r}(e^\Phi L_r)\right], \qquad (6.31)$$

where Q_ν and C_V denote the neutrino emissivity and the heat capacity, respectively. Integration of the above partial differential equations yields the time dependence of the temperature—that is, the cooling rate of the star—as well as the temperature profile in the star interior.

The cooling curves displayed in Fig. 6.4, corresponding to two neutron stars of mass 1.4 and 1.9 $M_\odot$ illustrate the critical role played by the onset of the direct Urca process in the more massive star, the central density of which, $\epsilon_c = 2.8 \times 10^{15}$ g cm^{-3}, exceeds the threshold discussed in Sect. 5.1.1. The two scenarios emerging from the figure are referred to as slow, or *standard*, and fast, or *enhanced*, cooling. Note that enhanced cooling can also be associated with the onset of direct Urca processes in strange baryonic matter or quark matter, discussed in, e.g., Ref. [107].

6.3 Tidal Deformation in Coalescing Binary Systems

A tide is the deformation of a body produced by the gravitational pull of another nearby body. Because the deformation depends on the body's internal structure, the observation of tidal effects in binary neutron star systems may provide valuable information on the EOS of neutron star matter.

The orbital motion of two stars gives rise to the emission of gravitational waves (GW), that carry away energy and angular momentum. This process leads to a decrease of the orbital radius and, conversely, to an increase of the orbital frequency.

In the early stage of the inspiral, characterised by large orbital separation and low frequency, the two stars—of mass M_1 and M_2, with $M_1 \geq M_2$—behave as point-like bodies and the evolution of the frequency is primarily determined by the chirp mass $\mathcal{M}$, defined as

$$\mathcal{M} = \frac{(M_1 M_2)^{3/5}}{(M_1 + M_2)^{1/5}} \, . \tag{6.32}$$

The details of the internal structure become important as the orbital separation approaches the size of the stars. The tidal field associated with one of the stars induces a mass-quadrupole moment on the companion, which in turn generates the same effect on the first star, thus accelerating coalescence. This effect is quantified by the tidal deformability, defined as

$$\Lambda = \frac{2}{3} k_2 \left(\frac{R}{GM} \right)^5 , \tag{6.33}$$

where M and R are the star's mass and radius, respectively, and k_2 is called second tidal *Love number* [108]. For any given stellar mass, the radius and the tidal Love number are uniquely determined by the EOS of neutron star matter.

According to the newtonian theory of gravity, the effect of a quadrupole tidal field is driven by the *tidal momentum*, defined as

$$\mathcal{E}_{ij} = -\left(\frac{\partial^2 \Phi}{\partial x_i \partial x_j} \right)_{\mathbf{x}=\mathbf{r}_c} , \tag{6.34}$$

where Φ is the external gravitational potential. The body subject to the tidal momentum, whose centre of mass position is denoted $\mathbf{r}_c$, develops a quadrupole deformation and the associated quadrupole moment

$$Q_{ij} = \int d^3 x \left(x_i x_j - \frac{1}{3} \delta_{ij} r \right) \varrho(\mathbf{x}), \tag{6.35}$$

where ϱ is the mass density and r is defined by the equation $r^2 = \delta_{ij} x_i x_j$.

The tensors Q_{ij} and $\mathcal{E}_{ij}$ are both symmetric and traceless. In the weak field approximation they are related through

$$Q_{ij} = -\Lambda \, \mathcal{E}_{ij}, \tag{6.36}$$

and simple dimensional considerations lead to the equation

$$\Lambda = \frac{2}{3} k_2 R^5 G^{-1} , \tag{6.37}$$

where the dimensionless constant k_2 is the second tidal Love number of Eq. (6.33), and 2/3 is a conventional factor.

The treatment of quadrupole deformations of neutron stars within general relativity involves the study of linearised perturbations of the equilibrium configurations [109]. The metric tensor is written in the form

$$g_{\alpha\beta} = g_{\alpha\beta}^{(0)} + h_{\alpha\beta} , \qquad (6.38)$$

where $g_{\alpha\beta}^{(0)}$ is the metric of a static and spherically-symmetric spacetime, given by Eq. (6.6), and the perturbation fulfils the requirement $|h_{\alpha\beta}| << 1$.

Quadrupole effects are associated with the $\ell = 2$ even-parity contribution to the expansion of $h_{\alpha\beta}$ in tensorial spherical harmonics, whose radial shape is described by the function $H(r)$ obeying the differential equation [109]

$$H'' + H' \left\{ \frac{2}{r} + e^{2\lambda} \left[\frac{2M(r)}{r^2} + 4\pi r(P - \epsilon) \right] \right\} \qquad (6.39)$$

$$+ H \left[-\frac{6e^{2\lambda}}{r^2} + 4\pi e^{2\lambda} \left(5\epsilon + 9P + \frac{\epsilon + P}{dP/d\epsilon} \right) - (2\nu')^2 \right] = 0 .$$

Integration of Eq. (6.39) and of the TOV equations (6.26) allows to determine the second tidal Love number, whose expression can be cast in the form

$$k_2 = \frac{8}{5} C^5 (1 - 2C)^2 [2 + 2C(y - 1) - y]$$

$$\times \left\{ 2C[6 - 3y + 3C(5y - 8)] + 4C^3[13 - 11y + C(3y - 2) \right. \qquad (6.40)$$

$$\left. + 2C^2(1 + y)] + 3(1 - 2C)^2[2 - y + 2C(y - 1)] \log(1 - 2C) \right\}^{-1} ,$$

with C and y given by

$$C = \frac{M}{R}, \quad y = R\frac{H'(R)}{H(R)} , \qquad (6.41)$$

where M and R denote the star mass and radius, respectively.

Equation (6.40) shows that, given a model of the EOS determining the values of M and R, a calculation of the tidal Love number k_2, requires the knowledge of the functions H and H', obtained from Eq. (6.39), evaluated at $r = R$.

Figure 6.5 illustrates the dependence of the tidal deformability, defined by Eq. (6.37), on the neutron star mass. The curve labelled N has been obtained using the EOSs of Ref. [63], corresponding to pure nucleonic matter, whereas the one labelled NY also includes hyperon contributions [110]. The emerging pattern shows that Λ has an appreciable sensitivity to the stiffness of the EOS. For a star of 1.4

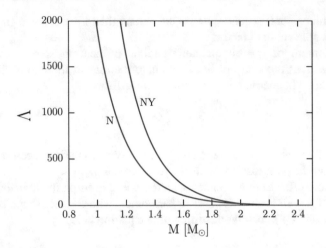

Fig. 6.5 Mass dependence of the tidal deformability defined by Eq. (6.37). The curves labelled N and NY have been obtained using the EOSs of Refs. [63] and [110], respectively. See text for details

solar masses, its value increases from $\Lambda \approx 260$, corresponding to the APR model, to $\Lambda \approx 600$ for the softer GM3 EOS of Glendenning and Moszkowski [110].

6.4   Neutron Star Oscillations

When a neutron star is perturbed by an external or internal event, it can be set into non-radial oscillations, leading to GW emission at the characteristic *complex* frequencies of its quasi-normal—that is, *damped*—modes (QNM).

The excitation of QNMs may be associated with, e.g., a glitch—that is, a sudden increase of the star rotational frequency—or a close interaction with an orbital companion, or a phase transition occurring in the neutron star core, or a gravitational collapse. The frequencies and damping times of the QNMs carry valuable information on both the structure of the star and the properties of matter in its interior. They can be computed by studying the adiabatic perturbations of an equilibrium configuration—which can be written as in Eq. (6.38)—with an assigned EOS, and solving the linearised Einstein equations, coupled to the equations of hydrodynamics, with suitably posed boundary conditions. For a more detailed discussion see, e.g., Ref. [104].

If the unperturbed star is assumed to be static and non rotating, it is convenient to expand all perturbed tensors in tensorial spherical harmonics, and since these harmonics have a different behaviour under the angular transformation $\theta \rightarrow \pi - \theta$ $\varphi \rightarrow \pi + \varphi$, the equations split into two decoupled sets: *polar* or *even*, belonging to parity $(-1)^{\ell}$, and *axial* or *odd*, belonging to parity $(-1)^{(\ell+1)}$. The polar equations are the relativistic generalization of the tidal perturbations of newtonian theory, and couple the perturbations of the gravitational field to perturbations of matter in the star interior.

As in Newton's theory, the classification of the polar modes is based on the source of the restoring force which prevails in bringing the perturbed fluid element back to the equilibrium position. For g-modes, or gravity modes, this force originates from a change of density, that is, from buoyancy, while for p-modes is due to a change of pressure. The *fundamental* mode, or f-mode, is the relativistic generalisation of the only possible oscillation mode of an incompressible homogeneous sphere. This classification scheme comprising g, f, and p modes was introduced in the 1940s. In general relativity, however, there exists an additional family of modes, dubbed w-modes, associated with pure space-time oscillations, with the corresponding motion of the fluid being negligible [111].

The frequency of the f-mode has been determined using several EOSs, including some obtained from highly realistic dynamical models. In addition, it has been shown that it is possible to infer empirical relations between the mode frequency and the macroscopic parameters specifying the star configuration, that is, mass and radius [112, 113]. The w-modes are characterised by frequencies typically higher than those of the g, f, and p modes, and much smaller damping times, implying that these modes are highly damped.

The imprint of the nuclear matter EOS of on QNMs can be illustrated considering, as a pedagogical example, the axial w-modes of a non rotating star. Unlike polar perturbations, the axial perturbations are not coupled to fluid motion, and do not have a Newtonian counterpart. Their radial dependence is described by a Schroödinger-like equation, involving a potential that depends explicitly on the distribution of energy and pressure in the interior of the star in the equilibrium configuration, that is, on the EOS. The explicit expression[2] is [114]

$$V_\ell(r) = \frac{e^{2v}}{r^3} \left\{ \ell(\ell+1)r + r^3\left[\epsilon(r) - P(r)\right] - 6M(r) \right\}, \tag{6.42}$$

with $M(r)$ defined as in Sect. 6.1, and

$$\frac{dv}{dr} = -\frac{1}{[\epsilon(r) + P(r)]}\frac{dP}{dr}. \tag{6.43}$$

It has been shown that, as in the case of the polar w-modes, it is possible to derive empirical relations that allow to determine the mass and radius of the star. The most remarkable result, however, is that, for any given value of the star compactness, M/R, the pulsation frequencies turn out to be ordered according to the stiffness of the EOS of matter in the star interior, which in turn determines the compressibility [115]. Higher frequencies always correspond to softer EOSs,

---

[2] Note that Eq. (6.42) is written using the geometrised unit system, in which $G = c = 1$.

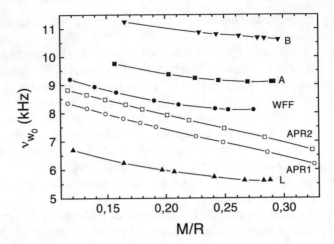

Fig. 6.6 Frequency of the first axial w-mode obtained from different models of the nuclear matter EOS—ordered according to stiffness—displayed as a function of the star compactness. The labels B and L correspond to the softest and stiffest EOSs, respectively. Reprinted from Ref. [115] with permissions, © RAS (1999). All rights reserved

regardless of the value of M/R. This pattern is displayed in Fig. 6.6, in which the stiffness of the EOS monotonically increases from model B to model L.

Note that, because the curves corresponding to different models do not cross each other within the range displayed in the figure, the results of Fig. 6.6 suggest that a measurement of the frequency of the first w-mode may allow to pin down the underlying EOS independent of the star compactness.

Part III

Multimessenger Neutron Star Astronomy

Observational Constraints on Theoretical Models

7

Abstract

Neutron stars have long been recognized as a potentially unparalleled source of information on strong interaction dynamics in the regime of high density and low temperature, which is not presently accessible in terrestrial laboratories. In principle, the data collected from neutron star observations may, in fact, set stringent constraints on theoretical models of neutron star matter. This chapter provides a brief overview of the available experimental information, as well as of the prospects for future developments.

7.1 The Golden Age of Neutron Stars

On August 17, 2017, the Advanced LIGO–Virgo detector network made the first observation of the GW signal labeled GW170817, consistent with emission from a coalescing binary neutron star system [116]. The detection of this signal, and the later observation of electromagnetic radiation by space- and ground-based telescopes [117] marked the dawning of the long anticipated age of multimessenger astronomy. A prominent role in this effort has been played the Neutron Star Internal Composition Explorer (NICER), deployed on the International Space Station. In view of the wealth of precise data that will be flowing from experimental facilities in the coming years, it has been argued that we are entering a *golden age* of neutron star physics, in which many outstanding and fundamental issues may be resolved [118].

Many ongoing studies are primarily aimed at exploiting the available and forthcoming empirical information to determine the occurrence of phase transitions, leading to the appearance of exotic forms of matter comprising strange baryons or deconfined quarks—possibly in a colour superconducting phase—in the inner core of the star. This chapter, however, will focus on the regime in which nucleons are

© The Author(s), under exclusive license to Springer Nature Switzerland AG 2023
O. Benhar, *Structure and Dynamics of Compact Stars*, Lecture Notes
in Physics 1019, https://doi.org/10.1007/978-3-031-35628-5_7

believed to be the relevant degrees of freedom, and nuclear matter theory is expected to be applicable for the description of neutron star properties.

The basic assumption underlying neutron star models based on nuclear theory is that, as long as the range of nuclear forces is negligible compared to the radius of spacetime curvature, the microscopic dynamics is totally unaffected from gravitational effects. It follows that under this condition nuclear interactions can be treated as if they were taking place in flat spacetime. In their classic book, Harrison et al. estimate the density at which the validity of this approximation breaks down to be $\varrho \sim 10^{49}$ g cm^{-3} [119].

7.2   Measurements of Mass and Radius

The connection between the EOS of neutron star matter and the star mass clearly emerged from the results of the study carried out by Oppenheimer and Volkoff in the 1930s, showing that the mass of a star consisting of non interacting neutrons is nearly constant for central densities $\varrho_c \gtrsim 2\varrho_0$, and cannot exceed $\sim$0.8 $M_\odot$ [18]. This findings rule out the description of nuclear matter based on the Fermi gas model, and, more generally, on any models predicting an EOS too soft to support stable stars with masses compatible with observations.

The recognition that the pressure responsible for neutron star equilibrium cannot be explained by Fermi-Dirac statistics alone, but is of primarily dynamical origin, has important phenomenological implications. It entails that experimental information on neutron star masses may be used to test the predictions of theoretical models of nuclear dynamics.

The masses of neutron stars belonging to binary systems have been determined to remarkable accuracy using a procedure based on Kepler's third law, describing the motion of two bodies of masses M_1 and M_2 orbiting about their center of mass [120]; see Fig. 7.1.

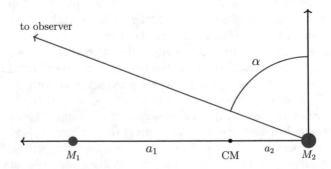

Fig. 7.1 Schematic representation of a binary system consisting of two stars of masses M_1 and M_2 orbiting about their center of mass CM

Consider a system consisting of a X-ray pulsar of mass M_X and an optical companion of mass M_{opt}. Kepler's law, states that

$$G\frac{G(M_X + M_{opt})}{a^3} = \left(\frac{2\pi}{T}\right)^2 , \tag{7.1}$$

where $a = a_x + a_{opt}$ and T is the orbital period. Combining the above relation with the equation yielding the projection of the orbital velocity of the X-ray pulsar along the line of sight

$$v_x = \frac{2\pi}{T} a_x \sin\alpha , \tag{7.2}$$

where α is the inclination with respect to the orbital plane, one obtains the mass function, defined as

$$f_X(M_X, M_{opt}, \alpha) = \frac{(M_{opt}\sin\alpha)^3}{(M_X + M_{opt})^2} = \frac{T}{2\pi G} v_x^3 , \tag{7.3}$$

which is a function of the measured orbital period and the orbital velocity, deduced by the observed Doppler shift of the X-ray pulse. The value of α depends on geometry, and in the case of eclipsing binaries the inclination is such that $\sin\alpha \approx 1$. If the mass functions of both the X-ray pulsar and the optical companion can be determined, the masses of the stars can be readily obtained from the relations

$$\left(\frac{f_{opt}}{f_X}\right)^{1/3} = \frac{M_X}{M_{opt}} = q , \quad M_X = f_X \frac{q(1+q^2)}{\sin^3\alpha} . \tag{7.4}$$

In addition to the X-ray binaries, there is a class of neutron star binaries known as radio pulsar binaries. Compared to X-ray pulsars, the pulsars belonging to these systems have shorter orbital and pulse period, and therefore allow for more precise timing measurements. The results of their observations provide information on *post-Keplerian* orbital parameters, describing effects of general relativity, such as the longitudinal advance of the periastron of the neutron star orbit, the time derivative of the period and the gravitational redshift, discussed in, e.g., Ref. [121].

A measurement of post-Keplerian parameters famously led to the precise determination of the masses of the neutron stars in the binary system PSR1913+16, discovered in 1974 by Hulse and Taylor [122]. Their analysis—that provided the first striking, although indirect, evidence of gravitational wave emission—yielded the results $M_1 = 1.39 \pm 0.15$ and $M_2 = 1.44 \pm 0.15\, M_\odot$ [123].

From the discussion of stellar equilibrium of Sect. 6.1, it follows that each EOS determines a specific value for the maximum mass of stable neutron star configurations, M_{max}. Hence, the most straightforward test of an EOS and the underlying microscopic model is a direct comparison between the corresponding M_{max} and the measured neutron star masses, summarised in Figs. 7.2 and 7.3. It is

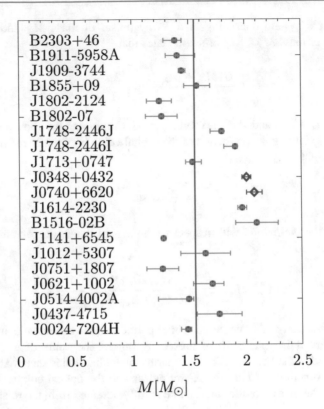

Fig. 7.2 Neutron star masses obtained from analyses of neutron star-white dwarf binaries. The dots display the data compiled by Kiziltan et al. [124], while the vertical line shows the central value of the corresponding distribution, $M = 1.55\ M_\odot$. The diamonds represent the masses of the pulsars J0348+0432 [125] and J0740+6620 [126]

apparent that, although the merged distribution is peaked around the canonical mass 1.4 $M_\odot$, the data points corresponding to the pulsars J1614-2230, J0348+0432 and J0740+6620, whose precisely determined masses turn out to be $M = 1.97 \pm 0.04$, 2.01 ± 0.04 and $2.08 \pm 0.07\ M_\odot$, respectively, conspicuously stick out.

The masses of the pulsars J1614-2230 and J0740+6620 have been determined combining measurements of the mass function and the relativistic effect known as Shapiro delay, the value of which determines the two additional parameters needed to obtain the masses of both stars in the binary system [126, 127]. In the case of the pulsar J0348+0432, on the other hand, the mass has been determined from the analysis of the spectral lines of its white dwarf companion [125].

Neutron star masses around and above 2 $M_\odot$ tend to rule out dynamical models predicting a soft EOS, notably those featuring the onset of a high-density phase including hyperons, discussed in Sect. 4.1. It should be kept in mind, however, that little is known about hyperon interactions in dense matter, and these models often imply rather crude assumptions. On the other hand, it is a fact that most models

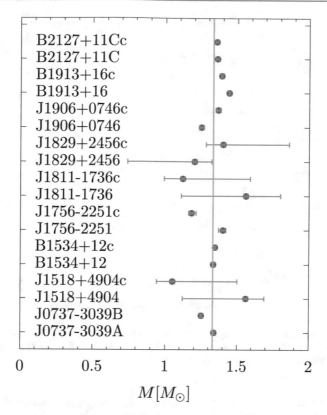

Fig. 7.3 Neutron star masses obtained from analyses of double neutron star binaries. The central value of the mass distribution, corresponding to $M = 1.33\ M_\odot$, is represented by the vertical line. Data taken from the compilation of Kiziltan et al. [124]

based only on nucleon degrees of freedom turn out to be compatible with the recent measurements.

The typical central-density dependence of the NS mass obtained from models involving nucleons only (stiff EOS), or nucleons and non interacting hyperons (soft EOS), is sketched in Fig. 7.4. The intermediate curve, obtained taking into account hyperon interactions, shows that their inclusion while resulting in a stiffer EOS and a larger value of M_{max}, falls conspicuously short of solving the *hyperon puzzle*, discussed in Sect. 4.1. As shown in Fig. 6.2, the qualitative pattern emerging from Fig. 7.4 is not modified by the effects of irreducible three-body interactions involving both nucleons and hyperons.

A detailed display of the results obtained from a purely nucleonic model is provided in Fig. 7.5, showing the neutron star masses obtained by Akmal et al. using different EOSs [63]. The meaning of the labels is same as in Figs. 3.9 and 3.10. It clearly appears that the inclusion of NNN interactions, described using either the UIX or the UIX$'$ potentials, is required to generate the pressure needed to support

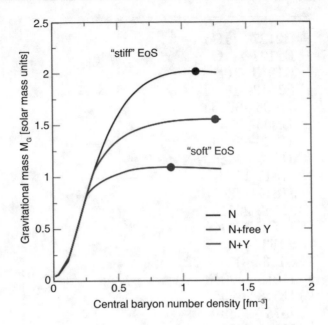

Fig. 7.4 Neutron star masses, in units of the solar mass, as a function of the star central density. The meaning of the curves is explained in the text. Reprinted from Ref. [128] under license CC-BY-3.0, © I. Vidaña

a stable neutron star of mass $\gtrsim 2$ $M_\odot$. However, as pointed out in Sect. 3.2, the stiffness of the EOS obtained from non relativistic approaches leads to a violation of causality at some large density $\bar{\varrho}$. The values of $\bar{\varrho}$ corresponding to the A18+UIX and A18+δv+UIX' Hamiltonians are indicated by the vertical bars of Fig. 7.5.

The inclusion of boost corrections appears to alleviate the violation of causality, pushing $\bar{\varrho}$ towards higher values. It is important to note, however, that the values of $\bar{\varrho}$ corresponding to both models including NNN interactions turn out to be larger than the central density of a neutron star of mass $M \sim 2$ $M_\odot$.

The overall picture emerging from the analyses of the available data—a summary of which can be found in, e.g., Ref. [129]—shows a significant degree of degeneracy. In fact, most EOSs derived from models of neutron star matter including only nucleons turn out to support a stable configuration of two solar masses. Precise measurements of the star radius, providing an additional constraint on the EOS and the underlying dynamical model, are needed to resolve this issue.

The radius of an ordinary star is obtained under the assumptions that

- the radiation emitted by the star can be described by a blackbody spectrum, which allows to determine the surface temperature T_s;
- the star radiates uniformly and isotropically.

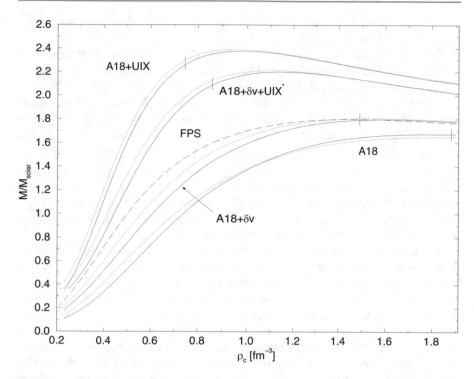

Fig. 7.5 Neutron star masses, expressed in units of the solar mass, as a function of central baryon number density. The curves have been obtained using different nuclear Hamiltonians. The values of ϱ corresponding to violation of causality are indicated by the vertical bars. The meaning of the labels is the same as in Figs. 3.9 and 3.10. Reprinted from Ref. [63] with permissions, © APS 1998. All rights reserved

Within this scheme, a measurement of flux and distance, denoted F and D, respectively, allows to determine the radius from

$$R = \sqrt{\frac{F D^2}{\sigma T_s^4}} , \tag{7.5}$$

where σ is the constant of Stephan-Boltzmann.

In the case of neutron stars, however, the models of spectrum and flux employed in data analysis involve uncertainties arising from astrophysical effects, that need to be carefully taken into account. Even more critical is the assumption that the whole surface of the star radiate uniformly. The emitted X-ray bursts actually originate from *hot spots*, first observed by the X-ray Multi-Mirror Mission (XMM-Newton) in the early 2000 [130]. More recently, the unprecedented capabilities of NICER's X-ray Timing Instrument (XTI), allowed to precisely track the rotation of the hot spots around the star, and detect the relativistic effects affecting their motion. These

observations have the potential to provide accurate information on both the mass
and radius of the emitting star.

Recent analyses of NICER data yielded the radius of a neutron star of mass $M = 1.4\ M_\odot$, the central density of which does not exceed $\sim 3\varrho_0$. The reported values—
$R = 12.45 \pm 0.65$ km [131], $12.33^{+0.76}_{-0.81}$ km [132], and $12.18^{+0.56}_{-0.79}$ km [133]—
turn out to be compatible with the predictions of theoretical calculations performed
using EOSs of purely nucleonic matter. The equatorial radius of the neutron star
J0740+6620, having mass $M = 2.072^{+0.067}_{-0.066}\ M_\odot$ and central density $\sim 4\varrho_0$, has
been also evaluated to be $R = 12.39^{+1.30}_{-0.98}$ km [132]. The small difference between
the radii of stars of mass 1.4 and $\sim 2.1\ M_\odot$, implying that the EOS is still quite
stiff at $\varrho > 3\varrho_0$, appears to rule out the occurrence of a strong first order phase
transition in the density range $3\varrho_0 \lesssim \varrho_B \lesssim 4\varrho_0$. However, It should be kept in mind
that—as pointed out in, e.g., Ref. [134]—a continuous crossover associated with the
formation of a mixed phase cannot be excluded.

In addition to the masses and tidal deformabilities, the data collected from the
groundbreaking observation of the gravitational wave event GW170817 have also
allowed to obtain the radii of the coalescing stars, although these results depend on
the EOS of neutron star matter employed in the analysis [135]. Figure 7.6 shows
a comparison between the mass-radius relations predicted by different models and
the region corresponding to the 90%-confidence-level estimates extracted from the
analysis of Ref. [135], represented by the box. The solid and dashed lines correspond
to the models of Akmal et al., discussed in Sect. 3.3, with (APR) and without (APR
NR) inclusion of relativistic boost corrections, respectively. The dotted line shows
the results of the GM3 model of Glendenning and Moszkowski [110], obtained
using the RMF formalism with a Lagrangian density including both nucleons and
hyperons. It appears that, although the accuracy of the data does not allow to fully

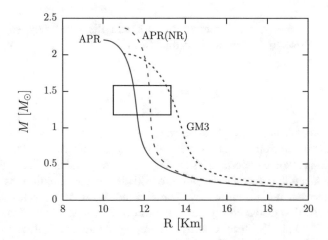

Fig. 7.6 The box represents the 90%-confidence-level estimate of the neutron star mass and radius
reported by the LIGO-Virgo Collaboration [135]. For comparison the solid, dashed and dotted lines
show the mass-radius relations predicted by the models described in the text

resolve the degeneracy between different models, the mass-radius relation predicted by the soft EOS of Ref. [110] appears to be only marginally compatible with observations.

Complementary information on neutron star radii have come from measurements of the gravitational redshift, z, related to the M/R ratio through

$$R(1 + z) = R \left(1 - 2G\frac{M}{R} \right)^{-1/2} . \tag{7.6}$$

In the early 2000s, Cottam et al. [136] reported the observation of spectroscopic lines associated with oxygen and iron transitions in the spectra of 28 bursts of the X-ray binary system EXO07948-676. The redshift value extracted from their analysis, $z = 0.35$, corresponding to $M/R = 0.153 \, M_\odot$/km, implies that the radii of neutron stars with masses in the range $1.4 \lesssim M \lesssim 1.8 \, M_\odot$ lie in the range $9 \lesssim R \lesssim 12$ km. The results of this and other similar studies further support the tendency to rule out soft EOS.

7.3   Measurements of the Tidal Deformability

The detection of the gravitational wave signal of event GW170817, reported by the LIGO/Virgo Collaboration in 2017 [116], allowed to obtain the chirp mass of the coalescing binary neutron star system, defined by Eq. (6.32), with extraordinary accuracy. The measured value turned out to be $\mathcal{M} = 1.188^{+0.004}_{-0.002} \, M_\odot$. On the other hand, the determination of the masses of the coalescing stars and their ratio, $q = M_2/M_1$ depends on the assumptions made on their spins, and involves larger uncertainty. The data of Ref. [116] have been analysed considering two different scenarios, corresponding to high and low spin. The results, yielding the range of both masses at 90% confidence level, are $1.36 \leq M_1 \leq 2.26 \, M_\odot$ and $0.86 \leq M_2 \leq 1.36 \, M_\odot$ for the high-spin scenario, and $1.36 \leq M_1 \leq 1.60 \, M_\odot$ and $1.17 \leq M_2 \leq 1.36 \, M_\odot$ for the low-spin scenario.

Figure 7.7 shows the two-dimensional probability density of the tidal deformabilities of the coalescing stars, Λ_1 and Λ_2, obtained from the analysis of the observation of the GW170817 event in the high spin scenario [116]. The boundaries of the regions enclosing 50% and 90% of the probability density are displayed by dashed lines. The predictions obtained from different models of the nuclear matter EOS are also shown, for comparison. The lines labelled APR and APR (NR) correspond to the EOSs of Ref. [63]—also described in Sect. 3.3—with and without the inclusion of relativistic boost corrections, respectively. The label GM3, on the other hand, refers to the model of Ref. [110]. The emerging pattern shows that the data favour EOSs predicting more compact stars. To see this, consider that the compactness of a star of 1.4 $M_\odot$ predicted by the APR, APR (NR) and GM3 models turns out to be $M/R = 0.108, 0.100$ and $0.091 \, M_\odot$/Km, respectively.

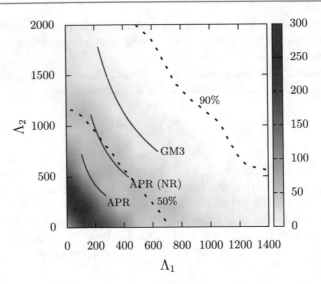

Fig. 7.7 Probability density of the tidal deformability parameters, Λ_1 and Λ_2, obtained from the analysis of the observation of the GW170817 event [116]. The thick solid lines represent the results of calculations carried out using the EOSs described in the text. The dashed lines show the boundaries of the regions enclosing 50% and 90% of the probability density. Adapted from Ref. [137] with permissions, © APS 2020. All rights reserved

7.4   Measurements of Neutron Star Cooling

As pointed out in Sect. 6.2, the cooling rate of a neutron star depends primarily on the onset of the direct Urca mechanism, which provides the dominant contribution to neutrino emissivity if the density of the central region exceeds the threshold discussed in Sect. 5.1.1. Depending on both its mass and the EOS of matter in its interior, the star is said to undergo enhanced cooling, if the direct Urca mechanics is active, or standard cooling, if neutrinos are mainly emitted through modified Urca processes. Significant modifications to this picture, however, may be arising from the presence of superfluid or superconducting phases in the neutron star core.

The possible emergence of an additional cooling mechanism below the critical temperature of the superfluid or superconducting transition, T_c, has been discussed by Page et al. [138, 139]. Within this scenario, referred to as *minimal* cooling, neutrino emission is significantly increased by the occurrence of Cooper pair formation and breaking. The temperature dependence of the corresponding emissivity turns out to follow the power law T^7, to be compared with the T^6 and T^8 dependences characteristic, respectively, of the direct and modified Urca processes; see Chap. 5.

The observed temperatures of neutron stars of different ages, summarised in Fig. 7.8, fail to provide a stringent test on the EOS and the underlying dynamical models, because most data points exhibit large error bars for both age and temperature, and the masses of the emitting stars are unknown. A more detailed information on the prevailing cooling mechanism has been provided by the observation of the

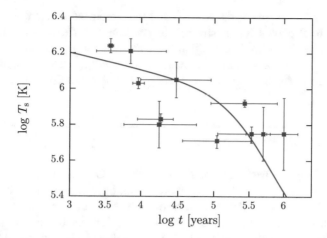

Fig. 7.8 Measured neutron star temperatures as a function of age. The experimental data are taken from Ref. [139]. For comparison the solid line illustrates the standard cooling evolution of a 1.4 $M_\odot$ neutron star predicted by the APR EOS of Ref. [63]

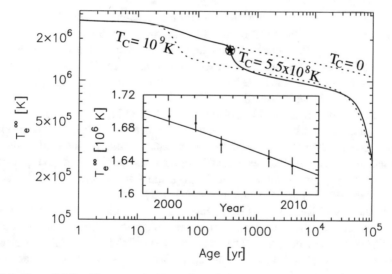

Fig. 7.9 The solid line illustrates the thermal evolution predicted by the intermediate cooling scenario for a 1.4 $M_\odot$ neutron star and critical temperatures $T_c = 5.5 \times 10^8$ K. For comparison, the dotted lines display the behaviour corresponding to $T_c = 0$—that is, in the absence of neutron superfluidity—and 10^9 K. The inset shows a comparison with the observed cooling of the neutron star in Cassiopea A [140, 141]. Reprinted from Ref. [142] with permissions, © APS 2011. All rights reserved

thermal evolution of the young neutron star in the supernova remnant Cassiopea A, whose age is estimated to be ~330 years.

Figure 7.9 shows that the minimal cooling scenario with critical temperature $T_c = 5.5 \times 10^8$ K provides a remarkably good description of the observed time evolution of the star temperature [140, 141]. The vertical axis of the figure gives

$T_e^\infty = (1 + z)T_e$, where z is the redshift and the effective temperature T_e is defined so that the total photon luminosity of the star can be written using the standard blackbody emission formula. Note that the data included in the figure are the only presently available measurement of neutron star cooling. The theoretical results correspond to a star of mass $M = 1.4M_\odot$, in which the direct Urca process is forbidden.

7.5   Towards Multimessenger Astronomy

Besides marking the beginning of the new era of GW astronomy, the landmark observation of event GW170817 contributed to highlight the potential of combining gravitational and electromagnetic observations. The association of the GW detection with that of the γ-ray burst GRB 170817A—carried out by the Fermi Gamma-ray Burst Monitor (GBM) [143] and the International Gamma-ray Astrophysics Laboratory (INTEGRAL) [144]—1.7 s after the coalescence, has been critical to confirm the hypothesis of neutron star merger, and provided the first direct evidence of the connection between these processes.

The observations of gravitational and electromagnetic signals can also be exploited to constrain the EOS or specific equilibrium properties of neutron star matter. Work along this line has been done to reconstruct the EOS within both phenomenological and nonparametric frameworks, determine the occurrence of phase transitions, or pin down the behaviour of the symmetry energy above nuclear saturation density [137, 145–170].

A somewhat different approach, discussed in Refs. [171, 172] aims at pushing the analyses based on multimessenger astrophysical data to a deeper level. The proposed strategy rests on the premise that the data currently available—as well as those to be collected by existing facilities operating at design sensitivity and next-generation detectors—may offer an unprecedented opportunity to constrain the microscopic models of nuclear dynamics at supranuclear density.

The analysis of Refs. [171, 172] focused on the repulsive three-nucleon interaction, which plays a critical role in determining the stiffness of the nuclear matter EOS at large densities. The strength of the NNN potential appearing in A18 + δv + UIX$'$ Hamiltonian, discussed in Sect. 3.3.5, was modified according to

$$V_{ijk}^R \rightarrow \alpha V_{ijk}^R \,, \tag{7.7}$$

and the value of the parameter α was inferred from a Bayesian analysis based on a dataset including

- the tidal deformabilities obtained from the observation of the GW170817 event performed by the LIGO/Virrgo Collaboration [116];
- the mass and radius of the millisecond pulsar PSR J0030+0451, measured by the NICER Collaboration [132];

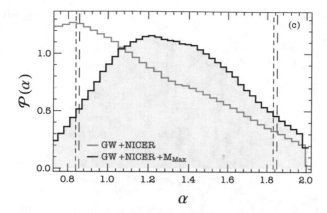

Fig. 7.10 The grey area represents the posterior probability distribution of the parameter α of Eq. (7.7), determining the strength of the repulsive three-nucleon force, inferred combining: (1) the GW observation of the binary system GW170817, (2) the mass radius constraints obtained by NICER for the millisecond pulsar PSR J0030+0451, and (3) the bound on the maximum neutron star mass imposed by PSR J0740+6620. Adapted from Ref. [171] with permissions, © APS 2021. All rights reserved

- the mass of PSR J0740+6620, the most massive pulsar discovered so far, having $M = 2.08^{+0.072}_{-0.069}\ M_\odot$ [126].

The resulting probability distribution is displayed in Fig. 7.10. While being rather broad, $\mathcal{P}(\alpha)$ exhibits a clear maximum, showing the sensitivity of multimessenger neutron star observations to the strength of repulsive three-nucleon forces. The position of the maximum, located at $\alpha > 1$, suggests that the data favour a potential stronger than the UIX model, which is recovered setting $\alpha = 1$. The corresponding EOS would be stiffer that the APR EOS at large density.

The prospects for obtaining more accurate information from future data—coming both from the achievement of design sensitivity of the LIGO/Virgo observatory and the advent of the proposed Einstein Telescope [173]—has been investigated in Ref. [172]. Figure 7.11 shows the probability distributions inferred from the detection of six simulated binary neutron star events by the Einstein Telescope. It is apparent that the distributions obtained setting $\alpha = 1$, and 1.3 can be unambiguously resolved.

The results reported in Refs. [171, 172] provide the first convincing evidence of the feasibility of exploiting multimessenger astrophysical data to obtain *direct* information on the dynamics of dense nuclear matter at microscopic level. Work along the line suggested in Refs. [171, 172] has been also carried out using the formalism based on the RMF approximation [174].

In view of the unprecedented success and the potential of multimessenger astronomy, the development of a unified model of nucleon systems from the deuteron to dense nuclear matter, which has long been regarded as the ultimate goal of nuclear theory, may be in fact achievable in the foreseeable future.

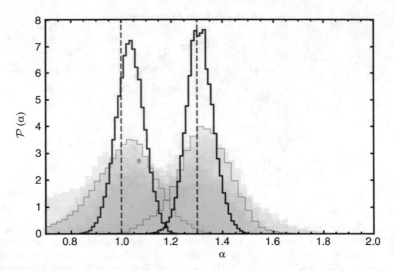

Fig. 7.11 The thick histograms represent the probability distributions inferred from the detection of six simulated binary neutron star events by the Einstein Telescope. The vertical dashed lines identify the values of α employed to generate the events. Reprinted from Ref. [172] with permissions, © APS 2022. All rights reserved

References

1. A.S. Eddington, *The Internal Constitution of the Stars* (Cambridge University Press, Cambridge, 1926)
2. S.E. Woosley, A. Heger, T.A. Weaver, Rev. Mod. Phys. **74**, 1015 (2002)
3. H.E. Bond et al., Astrophys. J. **840**, 70 (2017)
4. J.B. Holberg et al., Astrophys. J. **497**, 935 (1998)
5. J.L. Provencal et al., Astrophys. J. **568**, 324 (2002)
6. B.D. Mason et al., Astronom. J. **154**, 200 (2017)
7. J.L. Provencal et al., Astrophys. J. **494**, 759 (1998)
8. S. Chandrasekhar, Astrophys. J. **74**, 81 (1931)
9. L. Landau, E. Lifshiftz, *Statistical Physics* (Clarendon Press, Oxford, 1938)
10. J.H. Lane, Am. J. Sci. **50**, 148 (1870)
11. L. Rosenfeld, *Proceedings of the 16th Solvay Conefference on Astrophysics and Gravitation* (Edition de l'Université de Bruxelles, 1974)
12. L. Landau, Physik. Zeits. Sowjetunion **1**, 285 (1932)
13. D.G. Yakovlev, P. Haensel, G. Baym, C.J. Pethick, Physics–Uspekhi **56**, 289 (2013)
14. W. Baade, F. Zwicky, Proc. Natl. Acad. Sci. **20**, 255 (1934)
15. W. Baade, F. Zwicky, Phys. Rev. **46**, 76 (1934)
16. A. Hewish, S.J. Bell, J.D.H Pilkington, P.F. Scott, R.A. Collins, Nature **217**, 709 (1968)
17. R.C. Tolman, Phys. Rev. **55**, 364 (1939)
18. J.R. Oppenheimer, G.M. Volkoff, Phys. Rev. **55**, 374 (1939)
19. C. von Weizsäcker, Z. Phys. A **96**, 431 (1935)
20. G. Gamow, Proc. R. Soc. Lond. A **126**, 632 (1930)
21. G. Baym, C. Pethick, P. Sutherland, Astrophys. J. **170**, 299 (1971)
22. Facility of Rare Isotope Beams. https://frib.msu.edu/
23. Facility for Antiproton and Ion Research in Europe. https://fair-center.eu/
24. D.G. Ravenhall, C.J. Pethick, J.R. Wilson, Phys. Rev. Lett. **50**, 2066 (1983)
25. M. Okamoto, T. Maruyama, K. Yabana, T. Tatsumi, Phys. Rev. C **88**, 025801 (2013)
26. A.B. Migdal, JETP (Sov. Phys.) **10**, 176 (1960)
27. G. Baym, C. Pethick, D. Pines, Nature **224**, 673 (1969)
28. A. Sedrakian, J.W. Clark, Eur. Phys. J. A **167**, 55 (2019)
29. O. Benhar, G. De Rosi, J. Low Temp. Phys. **189**, 189 (2017)
30. A. Fabrocini, S. Fantoni, A. Yu. Illarionov, K.E. Schmidt, Phys. Rev. Lett. **95**, 192501 (2005)
31. N. Chamel, P. Haensel, Living Rev. Relativ. **11**, 10 (2008)
32. L. Kadanoff, G. Baym, *Quantum Statistical Mechanics* (Benjamin, Reading, 1976)
33. H. De Vries, C.W. De Jager, C. De Vries, At. Data Nucl. Data Tables **36**, 495 (1987)
34. S. Shlomo, V.M. Kolomietz, G. Colò, Eur. Phys. J. A **30**, 23 (2006)
35. G. Colò, Phys. Part. Nuclei **39**, 286 (2008)
36. Ya.B. Zel'dovich, Sov. Phys. JETP **14**, 1143 (1962)
37. H. Yukawa, Proc. Phys. Math. Soc. Jpn. **17**, 48 (1935)

© The Author(s), under exclusive license to Springer Nature Switzerland AG 2023
O. Benhar, *Structure and Dynamics of Compact Stars*, Lecture Notes
in Physics 1019, https://doi.org/10.1007/978-3-031-35628-5

38. J.D. Bjorken, S.D. Drell, *Relativistic Quantum Mechanics* (McGraw-Hill, New York, 1964)
39. N. Ishii, D. Aoki, T. Hatsuda, Phys. Rev. Lett. **99**, 022001 (2007)
40. R.B. Wiringa, V.G.J. Stoks, R. Schiavilla, Phys. Rev. C **51**, 38 (1995)
41. S. Weinberg, Phys. Lett. B **251**, 288 (1990)
42. E. Epelbaum, H.-W. Hammer, U.-G. Meißner, Rev. Mod. Phys. **81**, 1773 (2009)
43. R. Machleidt, D. Entem, Phys. Rep. **503**, 1 (2011)
44. I. Tews, J. Carlson, S. Gandolfi, S. Reddy, Astrophys. J. **860**, 149 (2018)
45. O. Benhar, Int. J. Mod. Phys. E **9** 2130009 (2021)
46. C. Hadjuk, P.U. Sauer, Nucl. Phys. A **322**, 329 (1979)
47. G.L. Payne, J.L. Friar, B.F. Gibson, I.R. Afnan, Phys. Rev. C **22**, 823 (1980)
48. B.S. Pudliner, V.R. Pandharipande, J. Carlson, R.B. Wiringa, Phys. Rev. Lett. **74**, 4396 (1995)
49. J. Fujita, H. Miyazawa, Prog. Theor. Phys. **17**, 3601957 (1957)
50. B.D. Day, J.S. Mc Carthy, T.W. Donnelly, I. Sick, Annu. Rev. Nucl. Part. Sci. **40**, 357 (1990)
51. J. Arrington et al., Phys. Rev. Lett. **82**, 2056 (1999)
52. R. Subedi et al., Science **320**, 1476 (2008)
53. J. Carlson, S. Gandolfi, F, Pederiva, S, Pieper, R. Schiavilla, K,E. Schmidt, R.B. Wiringa, Rev. Mod. Phys. **87**, 1067 (2015)
54. R.B. Wiringa, S.C. Pieper, J. Carlson, V.R. Pandharipande, Phys. Rev. C **62**, 014001 (2000)
55. S. Gandolfi, A. Lovato, J. Carlson, K.E. Schmidt, Phys. Rev. C **90**, 061306(R) (2014)
56. B. Povh, K. Rith, C. Scholz, F.Zetsche, W. Rodejohann, *Particles and Nuclei* (Springer, Heidelberg, 2014)
57. M. Baldo, in *Nuclear Matter and the Nuclear Equation of State*, ed. by M. Baldo (World Scientific, Singapore, 1990)
58. J.W. Clark, Prog. Part. Nucl. Phys. **2**, 89 (1979)
59. S. Fantoni, S. Rosati, Nuovo Cim. A **25**, 593 (1975)
60. V.R. Pandharipande, R.B. Wiringa, Rev. Mod. Phys. **51**, 821 (1979)
61. R.B. Wiringa, V. Fiks, A. Fabrocini, Phys. Rev. C **38**, 1010 (1988)
62. A. Akmal, V.R. Pandharipande, Phys. Rev. C **58**, 1804 (1998)
63. A. Akmal, V.R. Pandharipande, D.G. Ravenhall, Phys. Rev. C **56**, 2261 (1997)
64. I. Lagaris, V.R. Pandharipande, Nucl. Phys. A **359**, 349 (1981)
65. E. Feenberg, *Physics of Quantum Fluids* (Academic, New York, 1969)
66. J.L. Forest, V.R. Pandharipande, A. Arriaga, Phys. Rev. C **60**, 014002 (1999)
67. J.D. Walecka, Ann. Phys. **83**, 491 (1974)
68. A. Dadi, Phys. Rev. C **82**, 025203 (2010)
69. V.A. Ambartsumyan, G.S. Saakyan, Sov. Astron. **4**, 187 (1960)
70. S. Balberg, I. Lichtenstadt, G.B. Cook. Astrophys. J. Suppl. **121**, 515 (1999)
71. A. Gal, E.V. Hungerford, D.J. Millener, Rev. Mod. Phys. **88**, 035004 (2016)
72. S.A. Moszkowski, Phys. Rev. D **9**, 1613 (1974)
73. J. Haidenbauer, U.-G. Meißner, Phys. Rev. C **72**, 044005 (2005)
74. V.G.J. Stoks, Th.A. Rijken, Phys. Rev. C **59**, 3009 (1999)
75. H. Đapo, B.-J. Schaefer, J. Wambach, Phys. Rev. C **81**, 035803 (2010)
76. I. Vidaña, A. Polls, A. Ramos, L. Engvik, M. Hjorth-Jensen, Phys. Rev. **62**, 035801 (2000)
77. H. Polinder, J. Haidenbauer, U.-G. Meißner, Nucl. Phys. A **779**, 244 (2006)
78. H.-J. Schulze, A. Polls, A. Ramos, I. Vidaña, Phys. Rev. C **73**, 058801 (2006)
79. D. Lonardoni, A. Lovato, S. Gandolfi, F. Pederiva, Phys. Rev. Lett. **114**, 092301 (2015)
80. E. Lärmann, O. Philipsen, Ann. Rev. Nucl. Part. Sci **53**, 163 (2003)
81. F. Karsch, E. Lärmann, in *Quark-Gluon Plasma 3"*, ed. by R.C. Hwa, X.-N. Wang (World Scientific, Singapore, 2004)
82. J.I. Kapusta, *Finite-Temperature Field Theory* (Cambridge University Press, Cambridge, 1989)
83. M. Le Bellac, *Thermal Field Theory* (Cambridge University Press, Cambridge, 1996)
84. T. De Grand, R.L. Jaffe, K. Johnsson, J. Kiskis, Phys. Rev. D **12**, 2060 (1975)
85. G. Baym, S.A. Chin, Nucl. Phys. A **262**, 527 (1976)
86. R.L. Workman et al. (Particle Data Group), Prog. Theor. Exp. Phys. **2022**, 083C01 (2022)

87. N.K. Glendenning, Phys. Rev. D **46**, 1274 (1992)
88. G. Baym, S.A. Chin, Phys. Lett. B **62**, 241 (1976)
89. O. Benhar, R. Rubino, Astron. Astrophys. **434**, 247 (2005)
90. H. Heiselberg, C.J. Pethick, E.S. Staubo, Phys. Rev. Lett. **70**, 1355 (1993)
91. M.S. Berger, R.L. Jaffe, Phys. Rev. C **35**, 213 (1987). Erratum, *ibidem* C **44**, 566 (1991)
92. Y. Nambu, G. Jona-Lasinio, Phys. Rev. **122**, 345 (1961); *ibidem* **124**, 246 (1961)
93. M. Buballa, Phys. Rep. **407**, 205 (2005)
94. A.R. Bodmer, Phys. Rev. D **4**, 1601 (1971)
95. E. Witten, Phys. Rev. D **30**, 272 (1984)
96. G. Gamow, M. Shoenberg, Phys. Rev. **59**, 1117 (1940); *ibidem* **59**, 339 (1941)
97. G. Gamow, *My World Line: An Informal Autobiography* (The Viking Press, New York, 1970)
98. J. Boguta, Phys. Lett. B **106**, 255 (1981)
99. J.M. Lattimer, C.J. Pethick, M. Prakash, P. Haensel, Phys. Rev. Lett. **66**, 2701 (1991)
100. B.L. Friman, O.V. Maxwell, Astrophys. J. **232**, 541 (1979)
101. D.G. Yakovlev, A.D. Kaminker, O.Y. Gnedin, P. Haensel, Phys. Rep. **354**, 1 (2001)
102. D.G. Yakovlev, K.P. Levenfish, Astron. Astrophys. **297**, 717 (1995)
103. I.S. Gradshteyn, I.M. Ryzhik, *Table of Integrals, Series, and Products*, 7th edn. (Academic, New York, 2007)
104. V. Ferrari, L. Gualtieri, P. Pani, *General Relativity and Its Applications* (CRC Press, Boca Raton, 2021)
105. I. Vidaña, D. Logoteta, C. Providência, A. Polls, I. Bombaci, Europhys. Lett. **94**, 11002 (2011)
106. K.S. Thorne, Astrophys. J. **212**, 825 (1977)
107. C.J. Pethick, Rev. Mod. Phys. **64**, 1133 (1992)
108. T. Hinderer, Astrophys. J. **677**, 1216 (2008)
109. K.S. Thorne, A. Campolattaro, Astrophys. J. **149**, 591 (1967)
110. N.K. Glendenning, S.A. Moszkowski, Phys. Rev. Lett. **67**, 2414 (1991)
111. K.D. Kokkotas, B.F. Schutz, Mon. Not. R. Astron. Soc. **255**, 119 (1992)
112. N. Andersson, K.D. Kokkotas, Mon. Not. R. Astron. Soc. **299**, 1059 (1998)
113. O. Benhar, L. Gualtieri, V. Ferrari, Phys. Rev. D **70**, 124015 (2004)
114. S. Chandrasekhar, V. Ferrari, Proc. R. Soc. Lond. A **432**, 247 (1991)
115. O. Benhar, E. Berti, V. Ferrari, Mon. Not. R. Astron. Soc. **310**, 797 (1999)
116. B.P. Abbott et al., (LIGO Scientific Collaboration and Virgo Collaboration), Phys. Rev. Lett. **119**, 161101 (2017)
117. B.P. Abbott et al., (LIGO Scientific Collaboration and Virgo Collaboration), ApJ **848**, L12 (2017)
118. A. Mann, Nature **579**, 20 (2020)
119. B.K. Harrison, K.S. Thorne, M. Wakano, J. Wheeler, *Gravitation Theory and Gravitational Collapse* (University of Chicago Press, Chicago, 1965)
120. S.L. Shapiro, S.A. Teukolsky, *Black Holes, White Dwarfs, and Neutron Stars: The Physics of Compact Objects* (Wiley, New York, 1985)
121. T. Damour, N. Deruelle, Ann. Inst. Henri Poincaré, Phys. Theor. **4**, 263 (1986)
122. R.A. Hulse, J.H. Taylor, Astrophys. J. Lett. **195**, L51 (1975)
123. J.H. Taylor, L.A. Flower, P.M. McCulloch, Nature **277**, 437 (1979)
124. B. Kiziltan, A. Kottas, M. De Yoreo, S.E. Thorsett, Astrophys. J. **778**, 66 (2013)
125. J. Antoniadis et al., Science **340**, 6131 (2013)
126. E. Fonseca el al., Astrophys. J. Lett. **915**, L12 (2021)
127. P. Demorest, T. Pennucci, S. Ransom, M. Roberts, J. Hessels, Nature **467**, 1081 (2010)
128. I. Vidaña, J. Phys. Conf. Ser. **668**, 012031 (2016)
129. J.M. Lattimer, Universe **5**, 159 (2019)
130. A. De Luca et al., Astrophys. J. **623**, 1051 (2005)
131. M.C. Miller et al., Astrophys. J. Lett. **918**, L28 (2021)
132. T.E. Riley et al., Astrophys. J. Lett. **887**, L21 (2019)
133. G. Raaijmakers et al., Astrophys. J. Lett. **918**, L29 (2021)
134. L. Brandes, W. Weise, N. Kaiser, Phys. Rev. D **107**, 014011 (2023)

135. B.P. Abbott et al., (LIGO Scientific Collaboration and Virgo Collaboration), Phys. Rev. Lett. **121**, 161101 (2018)
136. J. Cottam, F. Paerels, M. Mendez, Nature **420**, 51 (2002)
137. A. Sabatucci, O. Benhar, Phys. Rev. C **101**, 045807 (2020)
138. D. Page, M. Prakash, J.M. Lattimer, A.W. Steiner, Proc. Sci. **XXXIV (BWNP)**, 005 (2011)
139. D. Page, J.M. Lattimer, M. Prakash, A.W. Steiner, Astrophys. J. Suppl. Ser. **155**, 623 (2004)
140. C.O. Heinke, A.C.G. Ho, Astrophys. J. Lett. **719**, L167 (2010)
141. P.S. Shternin, D.G. Yakovlev, C.O. Heinke, A.C.G. Ho, D.J. Patnaude, Mon. Not. R. Astron. Soc. **412**, L108 (2011)
142. D. Page, M. Prakash, J.M. Lattimer, A.W. Steiner, Phys. Rev. Lett. **106**, 081101 (2011)
143. A. Goldstein et al., Astrophys. J. Lett. **848**, L14 (2017)
144. V. Savchenko et al., Astrophys. J. Lett. **848**, L15 (2017)
145. E. Annala, T. Gorda, A. Kurkela, A. Vuorinen, Phys. Rev. Lett. **120**, 172703 (2018)
146. B. Margalit, B.D. Metzger, Astrophys. J. Lett. **850**, L19 (2017)
147. D. Radice, A. Perego, F. Zappa, S. Bernuzzi, Astrophys. J. Lett. **852**, L29 (2018)
148. A. Bauswein, O. Just, H.Th. Janka, N. Stergioulas, Astrophys. J. Lett. **850**, L34 (2017)
149. Y. Lim, J.W. Holt, Phys. Rev. Lett. **121**, 062701 (2018)
150. Y. Lim, A. Bhattacharya, J.W. Holt, D. Pati, Phys. Rev. C **104**, L032802 (2021)
151. E.R. Most, L.R. Weih, L. Rezzolla, J. Schaffner-Bielich, Phys. Rev. Lett. **120**, 261103 (2018)
152. S. De, D. Finstad, J.M. Lattimer, D.A. Brown, E. Berger, C.M. Biwer, Phys. Rev. Lett. **120**, 091102 (2018)
153. E. Annala, T. Gorda, A. Kurkela, J, Nättilä, A. Vuorinen, Nat. Phys. **16**, 907 (2020)
154. G. Raaijmakers et al., Astrophys. J. Lett. **893**, L21 (2020)
155. M.C. Miller, C. Chirenti, F.K. Lamb, Astrophys. J. **888**, 12 (2020)
156. B. Kumar, P. Landry, Phys. Rev. D **99**, 123026 (2019)
157. M. Fasano, T. Abdelsalhin, A. Maselli, V. Ferrari, Phys. Rev. Lett. **123**, 141101 (2019)
158. P. Landry, R. Essick, K. Chatziioannou, Phys. Rev. D **101**, 123007 (2020)
159. H. Güven, K. Bozkurt, E. Khan, J. Margueron, Phys. Rev. C **102**, 015805 (2020)
160. S. Traversi, P. Char, G. Pagliara, Astrophys. J. **897**, 165 (2020)
161. G. Raaijmakers et al., Astrophys. J. Lett. **918**, L29 (2021)
162. J. Zimmerman, Z. Carson, K. Schumacher, A.W. Steiner, K.Yagi, arXiv:2002.03210 [astro-ph.HE] (2020)
163. H.O. Silva, A.M. Holgado, A. Cárdenas-Avendaño, N. Yunes, Phys. Rev. Lett. **126**, 181101 (2021)
164. D. Blaschke, A. Ayriyan, D.E. Alvarez-Castillo, H. Grigorian, Universe **6**, 81 (2020)
165. S.-P. Tang, J.-L. Jiang, W.-H. Gao, Y.-Z. Fan, D.-M. Wei, Phys. Rev. D **103**, 063026 (2021)
166. B. Biswas, P. Char, R. Nandi, S. Bose, Phys. Rev. D **103**, 103015 (2021)
167. C. Pacilio, A. Maselli, M. Fasano, P. Pani, Phys. Rev. Lett. **128**, 101101 (2022)
168. T. Malik, C. Providência, Phys. Rev. D **106**, 063024 (2022)
169. S. Altiparmak, C. Ecker, L. Rezzolla, Astrophys. J. Lett. **939**, L34 (2022)
170. P.K. Gupta et al., arXiv:2205.01182 [gr-qc] (2022)
171. A. Maselli, A. Sabatucci, O. Benhar, Phys. Rev. C **103**, 065804 (2021)
172. A. Sabatucci, O. Benhar, A. Maselli, C. Pacilio, Phys. Rev. D **106**, 083010 (2022)
173. M. Maggiore et al., J. Cosmol. Astropart. Phys. **2020**, 050 (2020)
174. T. Malik, M. Ferreira, M. Bastos Albino, C. Providência, arXiv:2301.08169 [nucl-th] (2023)

Index

© The Author(s), under exclusive license to Springer Nature Switzerland AG 2023
O. Benhar, *Structure and Dynamics of Compact Stars*, Lecture Notes
in Physics 1019, https://doi.org/10.1007/978-3-031-35628-5

Printed in the United States
by Baker & Taylor Publisher Services